AF388983

A Themed Issue in Memory of Academician Zhu Yingguo (1939–2017)

A Themed Issue in Memory of Academician Zhu Yingguo (1939–2017)

Editors

Yuxian Zhu
Shaoqing Li
Guangcun He

Basel • Beijing • Wuhan • Barcelona • Belgrade • Novi Sad • Cluj • Manchester

Editors

Yuxian Zhu	Shaoqing Li	Guangcun He
Institute for Advanced Study	College of Life Sciences	College of Life Sciences
Wuhan University	Wuhan University	Wuhan University
Wuhan	Wuhan	Wuhan
China	China	China

Editorial Office
MDPI
St. Alban-Anlage 66
4052 Basel, Switzerland

This is a reprint of articles from the Special Issue published online in the open access journal *Agronomy* (ISSN 2073-4395) (available at: https://www.mdpi.com/journal/agronomy/special_issues/memory_academician_zhuyingguo).

For citation purposes, cite each article independently as indicated on the article page online and as indicated below:

Lastname, A.A.; Lastname, B.B. Article Title. *Journal Name* **Year**, *Volume Number*, Page Range.

ISBN 978-3-0365-9290-9 (Hbk)
ISBN 978-3-0365-9291-6 (PDF)
doi.org/10.3390/books978-3-0365-9291-6

Cover image courtesy of Jun Hu

Contents

About the Editors

Yuxian Zhu

Yuxian Zhu is an academician of the Chinese Academy of Sciences, an academician of the Academy of Sciences of Developing Countries, and a professor and doctoral supervisor at Wuhan University/Peking University. He is the Dean of the Institute for Advanced Study at Wuhan University, Vice Chairman of the Chinese Society of Plant Physiology and Molecular Biology, and Executive Deputy Editor-in-Chief of "Science China·Life Sciences." Zhu has published more than 130 peer-reviewed scientific research papers, of which more than 120 have been included in SCI. He is the author of the national common textbooks "Modern Molecular Biology" and "Experimental Technology of Molecular Biology," has presided over several major national scientific research projects, and has made outstanding contributions to the molecular mechanism of cotton fiber elongation and cotton genomics research. He won the second National Natural Science Award in 2011, the He Liang He Li Science and Technology Progress Award in 2012, and the second prize of the first National Teaching Materials Construction Award in 2021.

Shaoqing Li

Shaoqing Li is a professor and doctoral supervisor at Wuhan University and a Member of the Chinese Genetics Society and the Chinese Society of Plant Physiology. He is the director of the Key Laboratory of Indica Hybrid Vigor Research and Utilization of the Ministry of Agriculture. He has published over 90 SCI papers in internationally renowned journals, such as Nature Plants, Plant Cell, and New Phytologist, and he won the Hubei Province Science and Technology Progress Special Award and the Hubei Province Science and Technology Progress First Prize. The award, Hubei Province Outstanding Doctoral Dissertation, and other awards and honorary titles authorized seven national invention patents. He presided over projects such as the National Natural Science Foundation, the National Transgenic Project, the National 973, the National 863, and the Hubei Provincial Natural Fund key projects.

Guangcun He

Guangcun He is a professor and doctoral supervisor at Wuhan University. He has been engaged in research on rice genetics and brown planthopper resistance for a long time and has discovered and located multiple brown planthopper resistance genes from rice seed resources, cloned the first gene worldwide, deeply revealed the molecular mechanism of rice resistance to brown planthopper, and cultivated multiple resistance genes. Brown planthopper-resistant germplasm, genes, and technologies are widely used across the country, promoting the breeding and promotion of brown planthopper-resistant varieties. He has published more than 90 SCI papers in international academic journals, such as Nat Genet, PNAS, Mol Plant, and Plant Cell, and has obtained 15 authorized national invention patents and five international patents. He won the first prize in the Hubei Province Natural Science Award, the first prize in the Hubei Province Technology Invention Award, the special prize in the Dabeinong Science and Technology Award, and the second National Technology Invention Award prize.

Editorial

Rice Genomics Research, Gene Mining and Utilization: A Themed Issue Dedicated to Academician/Prof. Yingguo Zhu

Guangcun He [1], Shaoqing Li [1] and Yuxian Zhu [1,2,*]

[1] State Key Laboratory of Hybrid Rice, College of Life Sciences, Wuhan University, Wuhan 430072, China; gche@whu.edu.cn (G.H.); shaoqingli@whu.edu.cn (S.L.)

[2] Institute for Advanced Studies, Wuhan University, Wuhan 430072, China

* Correspondence: zhuyx@whu.edu.cn

Citation: He, G.; Li, S.; Zhu, Y. Rice Genomics Research, Gene Mining and Utilization: A Themed Issue Dedicated to Academician/Prof. Yingguo Zhu. *Agronomy* **2023**, *13*, 1015. https://doi.org/10.3390/agronomy13041015

Received: 9 March 2023

Accepted: 29 March 2023

Published: 30 March 2023

We are honored and privileged to edit this Special Issue, "Rice Genomics Research, Gene Mining and Utilization: A Themed Issue Dedicated to Academician Yingguo Zhu".

Rice is the primary staple food for over half of the world's population. Rice production has experienced several periods of rapid growth, referred to as green revolutions. The first such revolution was a consequence of the utilization of a semi-dwarf gene, while the second green revolution arose due to the development of hybrid rice, which has effectively and economically enhanced rice yield. In recent years, hybrid rice has come to cover 17 million hectares (ha), or 57%, of the total rice area in China, with a national average yield of 8 tons per hectare (t/ha); this is approximately 1.7 t/ha higher than the yield of conventional inbred varieties (6.3 t/ha). The annual increment in grain production in the country due to the cultivation of hybrid rice has the potential to feed an additional 70 million people per year. Therefore, hybrid rice has played, and continues to play, a crucial role in alleviating China's food scarcity problem and in ensuring that it is the largest self-sufficient country in terms of food. Professor Zhu is an outstanding rice geneticist both in China and worldwide. Throughout his 58-year scientific career, he has contributed to enriching the diversity of cultivated rice and the development of hybrid rice. His team successfully developed a Honglian-type cytoplasmic male sterile (HL-CMS) system in 1976 by crossing wild rice and cultivar rice. In 1985, he also designed the Maxie-type cytoplasmic male sterile (MX-CMS) system, by crossing the modern rice variety with a traditional landrace. These CMS systems have been used extensively in the development of hybrid rice in China and in the expansion of hybrid rice-growing areas in tropical Asian countries. Professor Zhu isolated the HL-CMS gene and the corresponding fertility-restoration genes, and elucidated the molecular mechanism. He has contributed to numerous aspects of hybrid rice genetics, including breeding and production. Professor Zhu has published more than 150 research papers, has co-authored four monographs, and has won both a second National Invention Award and a third National Natural Science Award.

Due to the growth in the human population, the need for sufficient and nutritious food has increased significantly in recent decades, while global climate change continues to threaten food production and specific aspects of food quality. With breakthroughs in hybrid rice research, the advent of high-quality, high-yield, multi-resistant rice that has wider adaptability, higher fertilization utility, requires limited pesticide application and has a higher stress tolerance is widely anticipated. In this Special Issue, we are pleased to publish ten papers attend to various aspects of this research area, including rice genomics, gene mining, and cultivation towards the development of the ideal hybrid rice.

Cultivated rice *O. sativa* is divided into two subspecies, indica and japonica, which differ widely, in genomics and in their morphological, agronomical, and ecological traits. Transposons (TEs) are noted for their ability to alter gene expression and function, and are the cause of plant speciation and evolution. Professor Yuxian Zhu and his colleagues developed a method named matrix-TE, in order to comprehensively investigate the differentiation of intact and truncated LTR/TEs in indica and japonica genomes [1]. As such, six LTR/TE superfamilies were identified. They observed that the indica rice-specific TE peak is P-Gypsy and that the japonica rice-specific TE peak is P-Copia. The single TE peak P-Gypsy was observed in centromeric regions of the indica genome. By utilizing the matrix TE method, the divergence in the indica and japonica genomes, especially their centromeric regions, was discovered to primarily be a consequence of Ty3/Gypsy insertions at 0.77 Mya. Inter-subspecific hybrids between indica and japonica possess enhanced heterosis, and often produce greater yields compared to intra-subspecific hybrids. However, the sterility of inter-subspecific hybrids prevents the application of inter-subspecific heterosis. Wide compatibility varieties (WCVs) permit interspecific hybrids to achieve normal fertility. Zhao et al. reported upon the *F12* gene, which affected pollen and spikelet fertility, and improved the performance of the indica–japonica hybrid [2]. The *F12* gene was mapped to a region of 630 kb, and was flanked by the D1101 and D1164 markers on chromosome 12. In this region, two putative genes were predicted to be candidates for the wide compatibility genes (WCGs), and deletions/insertions within the exons of both putative genes were observed between indica and japonica rice. Identifying *F12* thus provides more opportunities for the further exploitation of inter-subspecific hybrids in rice. This discovery will also aid the elucidation of the mechanism involved in the wide compatibility and development of inter-subspecific hybrid rice.

Deng et al. developed a RIL population and applied testcrosses that were derived by crossing RILs with two cytoplasmic male sterile lines; this was performed in order to dissect heterosis loci and thus report six heterosis-related QTLs of panicle numbers (PN) and five heterosis-related QTLs of tiller numbers (TN), including several known genes: *MOC1*, *TAC1* and *OsETR2* [3]. They concluded that several serine/threonine protein kinase genes may play an important role in heterosis, which encourages the production of a high yield. Chlorophyll biosynthesis and chloroplast development affect photosynthetic activity, thus regulating grain production. Lu et al. generated a yellow-green-leaf mutant named ygl9311 [4]. Compared with the wild type, the content of the photosynthetic pigment in ygl9311 leaves was significantly reduced, and chloroplast development was delayed. They mapped the mutant gene to a 430 kb region on chromosome 3. They performed transcriptome sequencing analysis and revealed that the candidate gene, *OsChlC1* (BGIOSGA012976), encodes a Mg-chelatase I subunit. The CRISPR/Cas9 knockout of *OsChlC1* reproduced the same yellow-green leaf phenotype, revealing that *OsChlC1* is plays an essential role in maintaining chlorophyll content and in aiding chloroplast development in rice leaves.

The heading date is a fundamental factor in the determination of crop yield. Yuan et al. isolated an early-flowering gene *OsHd8* on the short arm of chromosome 8 in the early-flowering rice JiaHong2B (JH2B) through map-based cloning [5]. *OsHd8* encodes a putative HAP3 subunit of the CCAAT-box-binding transcription factor and regulates the expression of *OsGI*, *OsSDG718*, *OsHDT1*, and *OsGHD7.1* in order to promote rice heading under long-day conditions. A genetic divergence analysis at the *OsHd8* locus exhibited strong genetic differentiation between the indica and japonica subspecies, suggestive of artificial selection during the domestication of cultivated rice. Gibberellins (GAs) are a large group of diterpene plant hormones that play essential roles in rice development processes, such as seed germination, stem elongation, root development, pollen development, and flower induction. Gibberellin-dioxygenases (GAoxes) are involved in the biosynthesis and deactivation of gibberellins. He et al. conducted a comprehensive genome-wide investigation of GA oxidases in rice [6]. They identified 80 candidate *OsGAox* genes, of which 19 were further analyzed. RNA-seq data indicated that all GAox genes exhibited tissue-specific expression patterns in the leaf, shoot apical meristem, inflorescence, and seed. GA_3 treatment resulted in the differential expression of *OsGAox* genes. The chalkiness of rice is characterized by the opaque part of the grain, which raises the incidence of grain breakage and reduces its milling yield. Together with the white core (center), these characteristics are considered critical factors regarding rice quality and its commercial value. In the study by Shi et al., a QTL of white-core rate (WCR), *qWCR4*, was fine-mapped to a 35 kb region containing six annotated genes on chromosome 4, using a BC_5F_2 population [7]. Based on the relative expression levels of these genes in different endosperm developmental stages and on nucleotide diversities of the two parents, LOC_Os04g50060 and LOC_Os04g50070 in the *qWCR4* region were identified as the candidate genes for WCR. Compared with NIL (BL130), which has a lower WCR, starch granules in the central endosperm of chalky grains of NIL (J23B) displayed higher WCR values; they also possessed a typical round and loosely packed morphology, as well as a higher rate of seed filling. These results pave the way for improving rice yield and grain quality.

Brown planthopper (BPH) is rice's most devastating insect pest and is a severe threat to rice production. Exploring new resistance genes and integrating them into the rice genome is necessary for a durable BPH-resistant variety. Kim et al. identified and mapped the BPH-resistance gene *Bph43* to a region of ~380 kb on chromosome 11 [8]. A gene cluster that encodes putative nucleotide-binding domain leucine-rich repeat-containing (NBS-LRR) proteins and LRR family proteins was identified in the *Bph43* region. A highly resistant near-isogenic line (NILBph43-9311) was developed by introgressing *Bph43* into the elite restorer line 9311 through marker-assisted selection. NIL-Bph43-9311 conferred strong antibiosis and antixenosis effects on BPH. A comparative transcriptome analysis revealed the presence of 194 upregulated and 183 downregulated genes in BPH-infested NIL-Bph43-9311.

Lodging severely reduces the grain yield and quality of rice. In order to decipher the mechanisms underlying the lodging response to temperature and solar radiation, Luo et al. analyzed the lodging resistance of 12 indica rice varieties at two eco-sites on three sowing dates for three consecutive years [9]. They demonstrated that temperature had a negative effect, while solar radiation positively affected the lodging resistance. A high temperature resulted in a low lodging resistance, since it reduced the culm's physical strength by producing a longer and thinner basal second internode. The variety Chuanxiang 29B was most sensitive to temperature, and the lodging-resistant cultivar Jiangan was least responsive to temperature. In traditional rice production, the rice seeds are sown in a nursery to raise rice seedling rates, and then the rice seedlings are manually transplanted to the puddled field, which involves strenuous manual labor. Thus, developing an efficient and high-yielding mechanical rice establishment system is of significant value in the quest for efficient and large-scale rice production. Wang et al. conducted a two-year field experiment using the orderly mechanical rice seedling throwing system (OMST) [10]. The grain yield with the OMST system was significantly enhanced compared to the manual seedling throwing (MST) method, and was equivalent to the manual transplanting (MT) method. The final tiller number, panicle number, and total spikelet number, as well as the final yield when utilizing the OMST system, were significantly elevated compared to that of the MST method. Their study suggests that the OMST system is an efficient and high-yielding rice establishment method that possesses the potential to replace traditional manual transplanting methods in rice production. All of these excellent works contribute significantly to the development of hybrid rice. We would like to take this opportunity to thank all the contributors to this Special Issue.

Author Contributions: G.H. prepared a draft and S.L. and Y.Z. reviewed and edited for clearness. All authors have read and agreed to the published version of the manuscript.

Funding: This research received no external funding.

Conflicts of Interest: The authors declare no conflict of interest.

References

1. Wu, Z.; Xi, W.; Han, Z.; Wu, Y.; Guan, Y.; Zhu, Y. Genome-Wide Comparative Analysis of Transposable Elements by Matrix-TE Method Revealed Indica and Japonica Rice Evolution. *Agronomy* **2022**, *12*, 1490. [CrossRef]
2. Zhao, W.; Zhou, W.; Geng, H.; Fu, J.; Dan, Z.; Zeng, Y.; Xu, W.; Hu, Z.; Huang, W. Identification of a New Wide-Compatibility Locus in Inter-Subspecific Hybrids of Rice (*Oryza sativa* L.). *Agronomy* **2022**, *12*, 2851. [CrossRef]

3. Deng, X.; Wang, J.; Liu, X.; Yang, J.; Zhou, M.; Kong, W.; Jiang, Y.; Ke, S.; Sun, T.; Li, Y. QTL Analysis and Heterosis Loci of Effective Tiller Using Three Genetic Populations Derived from Indica-Japonica Crosses in Rice. *Agronomy* **2022**, *12*, 2171. [CrossRef]
4. Lu, W.; Teng, Y.; He, F.; Wang, X.; Qin, Y.; Cheng, G.; Xu, X.; Wang, C.; Tan, Y. *OsChlC1*, a Novel Gene Encoding Magnesium-Chelating Enzyme, Affects the Content of Chlorophyll in Rice. *Agronomy* **2023**, *13*, 129. [CrossRef]
5. Yuan, H.; Wang, R.; Cheng, M.; Wei, X.; Wang, W.; Fan, F.; Zhang, L.; Wang, Z.; Tian, Z.; Li, S. Natural Variation of *OsHd8* Regulates Heading Date in Rice. *Agronomy* **2022**, *12*, 2260. [CrossRef]
6. He, Y.; Liu, W.; Huang, Z.; Huang, J.; Xu, Y.; Zhang, Q.; Hu, J. Genome-Wide Analysis of the Rice Gibberellin Dioxygenases Family Genes. *Agronomy* **2022**, *12*, 1627. [CrossRef]
7. Shi, H.; Zhu, Y.; Yun, P.; Lou, G.; Wang, L.; Wang, Y.; Gao, G.; Zhang, Q.; Li, X.; He, Y. Fine Mapping of *qWCR4*, a Rice Chalkiness QTL Affecting Yield and Quality. *Agronomy* **2022**, *12*, 706. [CrossRef]
8. Kim, J.; An, X.; Yang, K.; Miao, S.; Qin, Y.; Hu, Y.; Du, B.; Zhu, L.; He, G.; Chen, R. Molecular Mapping of a New Brown Planthopper Resistance Gene *Bph43* in Rice (*Oryza sativa* L.). *Agronomy* **2022**, *12*, 808. [CrossRef]
9. Luo, X.; Wu, Z.; Fu, L.; Dan, Z.; Long, W.; Yuan, Z.; Liang, T.; Zhu, R.; Hu, Z.; Wu, X. Responses of the Lodging Resistance of Indica Rice Cultivars to Temperature and Solar Radiation under Field Conditions. *Agronomy* **2022**, *12*, 2603. [CrossRef]
10. Wang, W.; Xiang, L.; Zheng, H.; Tang, Q. Orderly Mechanical Seedling-Throwing: An Efficient and High Yielding Establishment Method for Rice Production. *Agronomy* **2022**, *12*, 2837. [CrossRef]

Obituary

Brief Biography of Professor Yingguo Zhu

Jun Hu [1] and Yuxian Zhu [1,2,*]

[1] State Key Laboratory of Hybrid Rice, Engineering Research Center for Plant Biotechnology and Germplasm Utilization of Ministry of Education, College of Life Sciences, Wuhan University, Wuhan 430072, China; junhu@whu.edu.cn

[2] Institute for Advanced Studies, Wuhan University, Wuhan 430072, China

* Correspondence: zhuyx@whu.edu.cn

Citation: Hu, J.; Zhu, Y. Brief Biography of Professor Yingguo Zhu. *Agronomy* **2022**, *12*, 858. https://doi.org/10.3390/agronomy12040858

Academic Editor: Xinghua Wei

Received: 15 March 2022
Accepted: 28 March 2022
Published: 31 March 2022

Publisher's Note: MDPI stays neutral with regard to jurisdictional claims in published maps and institutional affiliations.

Professor Yingguo Zhu was born in November 1939 and grew up in a poor mountain village in Hubei Province, China. Since his early childhood was rooted in two major global wars, especially in China, his ordinary farmer parents had great difficulty raising him and their two other children. As the eldest son, he was used to the responsibility of caring for his little brother and sister at a very young age. He exhibited abundant talent as early as elementary school in Luotian village. In 1956, it was recommended that he study in Luotian's number one high school. Three years later, he was enrolled at Wuhan University after achieving excellent grades. During the "Three Years of Natural Disasters" (1959–1962), he witnessed a huge amount of death by hunger among his fellow peasants, friends and relatives. This horrible situation made him determined to devote his life to the field of global food security.

Prof. Yingguo Zhu studied and worked at Wuhan University for 58 years and contributed to enriching the diversity of hybrid rice. He pursued his dream of creating new rice germplasms by crossing different cultivar varieties with wild rice. Starting in 1964, he collected numerous rice varieties, and put all his time and energy into hybrid rice, including three-line and two-line breeding. He realized very early on that the creation of male sterile lines was a priority for hybrid rice. Owing to his great efforts, Honglian-type cytoplasmic male sterile (HL-CMS) lines were successfully bred in 1976, and Maxie-type cytoplasmic male sterile (MX-CMS) lines were bred in 1985. HL-CMS was derived from the crossing of wild rice and a cultivar rice, whereas MX-CMS was derived from the crossing of two cultivar rice varieties. From these data, Prof. Zhu suggested that CMS genes were widely spread in rice varieties. Subsequently, his team cloned the HL-CMS gene, *orfH79*, which is located downstream of *atp6* in the mitochondrial genome. They also cloned two fertility-restoration genes, *Rf5* and *Rf6*, for HL-CMS. These genes controlled HL-type rice hybridization with high efficiency. Meanwhile, his team also worked on two-line hybrid rice from 1986, and bred several thermo-sensitive male sterile lines. Many hybrid rice varieties bred by his team have been planted in China, and have also been released to several other countries in south-east Asia and Africa. Until 2021, it was estimated that those hybrid rice combinations have been planted in over 10 million hectares worldwide, and have made great contributions to improving food supply in relatively poor countries. The utilization of various rice hybrids has improved not only the yields of rice, but also the incomes of farmers. Thus, he realized his childhood wish, and the world will remember him for his contributions to food security.

Working as an ordinary teacher at Wuhan University, Prof. Zhu has taught many students in his 58-year career. He was modest, decent and polite, and always instructed his students through his words and deeds. He encouraged them to study new technologies with an open mind, and he wanted them to work hard to serve their motherland. His team has published more than 50 articles and made key contributions to hybrid rice research. He loved his family and his country deeply, and showed great passions for reading books and walking on the rice field banks. In 2015, he began to suffer from myelodysplastic syndrome, yet he persisted in working in the laboratory, as well as in the rice field. He finally succumbed to the disease on 9 August 2017, at the relatively young age of 79.

"Struggle and strive until you die" were among the last words he left to his students. As always, he insisted on encouraging people, especially his students, to work hard for food security around the world.

Author Contributions: Writing—original draft preparation, J.H.; writing—review and editing, Y.Z. All authors have read and agreed to the published version of the manuscript.

Conflicts of Interest: The authors declare no conflict of interest.

Article

Fine Mapping of *qWCR4*, a Rice Chalkiness QTL Affecting Yield and Quality

Huan Shi, Yun Zhu, Peng Yun, Guangming Lou, Lu Wang, Yipei Wang, Guanjun Gao *, Qinglu Zhang, Xianghua Li and Yuqing He *

National Key Laboratory of Crop Genetic Improvement, Hubei Hongshan Laboratory, Huazhong Agricultural University, Wuhan 430070, China; huanshi1023@gmail.com (H.S.); zy9000722@gmail.com (Y.Z.); pengyun0106@gmail.com (P.Y.); louguangming@mail.hzau.edu.cn (G.L.); wanglu001222@gmail.com (L.W.); wangyipei3@gmail.com (Y.W.); qingluzhang@mail.hzau.edu.cn (Q.Z.); xhli@mail.hzau.edu.cn (X.L.)
* Correspondence: gaojun8199@webmail.hzau.edu.cn (G.G.); yqhe@mail.hzau.edu.cn (Y.H.)

Abstract: Rice (*Oryza sativa* L.) chalkiness greatly reduces the rice quality and the commercial value. In this study, *qWCR4*, a previously reported quantitative trait locus (QTL) of white-core rate (WCR), was confirmed by a BC_5F_2 segregation population and further fine mapped to a 35.26 kb region. In the *qWCR4* region, *LOC_Os04g50060* and *LOC_Os04g50070* showed significant differences in expression level in endosperm between two NILs, whereas four other genes had no expression. Starch granules in the central endosperm of chalky grains from NIL(J23B) with higher WCR exhibited a typically round and loosely packed morphology. NIL(J23B) with higher WCR accompanied a higher seed filling speed. Moreover, $qWCR4^{J23B}$ (*qWCR4* allele in J23B) increased WCR, grain numbers per plant, seed setting rate, grain width, and thousand-grain weight, contributing to a superior yield per plant. All in all, our research results not only lay a foundation for map-based cloning of *qWCR4* but also provide new genetic resources for rice yield and quality breeding.

Keywords: rice; white-core rate; fine-mapping; quantitative RT-PCR; yield; and quality

Citation: Shi, H.; Zhu, Y.; Yun, P.; Lou, G.; Wang, L.; Wang, Y.; Gao, G.; Zhang, Q.; Li, X.; He, Y. Fine Mapping of *qWCR4*, a Rice Chalkiness QTL Affecting Yield and Quality. *Agronomy* **2022**, *12*, 706. https://doi.org/10.3390/agronomy12030706

Academic Editor: HongWei Cai

Received: 7 February 2022
Accepted: 10 March 2022
Published: 14 March 2022

Publisher's Note: MDPI stays neutral with regard to jurisdictional claims in published maps and institutional affiliations.

1. Introduction

Rice (*Oryza sativa* L.) is a primary food crop that has greatly contributed to solving the world food security crisis. The rice yield has been improved dramatically after the first and second rice green revolutions in the past several decades [1]. In recent years, more and more attention has focused on the quality of rice concerning the improvement of people's living standards. Rice chalkiness, as an extremely unsatisfactory quality trait in rice sales and consumption, not only affects grain appearance but also has negative effects on rice processing and cooking characteristics [2].

The chalkiness of rice is characterized by the opaque part of the grain, which can be divided into white-belly (ventral side), white-core (center), and white-back (dorsal side) according to the opaque position [3]. The opaque endosperm is often associated with changes in the morphology and arrangement of starch granules, characterized by round and loosely packed starch granules [4–7]. Many studies have shown that chalkiness affects heading yield and milling yield by increasing the incidence of grain breakage [4,5,8]. The chalky grains exhibited lower protein content, and the palatability of cooked rice showed a linear decrease with increasing chalky rice proportion in the sensory evaluation [6]. Moreover, the chalkiness of kernels is often accompanied by low amylose content, which influences the sensory quality [9]. Thus, reducing the chalkiness rate has become one of the important goals in high-quality rice breeding.

Rice chalkiness is affected by genetic and environmental factors. Rice endosperm filling is a continuous process that typically lasts for more than a month. In this process, abiotic stresses, especially high temperature, could accelerate seed filling and led to chalkiness formation. Under high temperature conditions, chalkiness is usually triggered within 9–16 days after pollination [10]. Reporters have pointed out that warmer weather during the filling stage reduces grain appearance and palatability due to chalkiness in rice grains [8,11,12]. Chen et al. [13] found that high temperature could cause the increase of amylase and the decrease of GBSS, SBEs, and BEIIb activities, and these changes are the direct or indirect reasons for the chalkiness. Therefore, high temperature is the main abiotic stress leading to the increase of rice chalkiness. Genetic background is another determinant of rice chalkiness. Grain chalkiness is a complex quantitative trait, and more than one hundred quantitative trait loci (QTLs) for rice chalkiness have been reported on all 12 chromosomes [10–12,14–17]. In addition, rice grain shape is another determinant of rice chalkiness. Changes in grain size are usually accompanied by changes in chalkiness [18,19]. Floury mutants are valuable genetic resources for dissecting the underlying mechanisms of chalkiness formation. So far, many genes responsible for floury phenotype have been confirmed in directly influenced starch and protein synthesis, and presented an extremely chalky phenotype [20–25]. However, only a few QTLs for chalkiness were successfully fine-mapped or cloned [26–28]. Among them, *Chalk5* [26] encodes a vacuolar H^+-translocating pyrophosphatase. The elevated expression of *Chalk5* increased the chalkiness degree of endosperm, which may be due to the disruption of the pH dynamic balance of the endosperm transport system during seed development, thus affecting the biogenesis of protein bodies, and accompanied by the increase of vesicle structure, thus forming gaps between endosperm storage substances and resulting in the generation of chalky grains. *WCR1* [28] encodes an F-box protein that negatively regulates grain chalkiness. *WCR1* promotes the elimination of excess ROS, maintains redox homeostasis, and delays programmed cell death in starch endosperm by up-regulating the transcription of *MT2b*. Hence, the discovery of more chalkiness-related genes will be helpful to the revelation of the chalkiness genetic mechanism.

In this study, a stable QTL for white-core rate in rice, *qWCR4*, was confirmed and fine mapped to a 35.26 kb region. RT-qPCR and parental sequence analysis results showed that *LOC_Os04g50060* and *LOC_Os04g50070* could be the causal genes for *qWCR4*. NIL(J23B) with higher WCR was accompanied by faster seed filling speed, bigger grain size, higher yield per plant, and amylose content. In summary, the knowledge will pave the way for improving rice yield and quality.

2. Materials and Methods

2.1. Plant Materials and Field Experiment

In a previous study, an RIL population consisting of 184 lines was developed from a cross between an *indica* cultivar J23B (the recurrent parent) and a *japonica* cultivar BL130 (the donor parent) [17]. *qWCR4* was repeatedly detected to affect the white-core rate in two environments by QTL mapping. To find the underlying gene controlling WCR in the *qWCR4* interval, we selected one F_7 generation RIL containing the BL130 alleles of *qWCR4* to backcross with the recipient parent J23B. Five times backcrosses were performed for *qWCR4*, and the two flanking markers of the QTL mapping interval were used for hybrids screening for authenticity. After obtaining the BC_5F_1 hybrids, plants with a heterozygous fragment of the QTL mapping interval were retained and self-pollinated to obtain the BC_5F_2 population defined as the NIL-F_2 population. A total of 192 individuals were planted in 16 rows of 12 plants each for QTL validation.

For fine mapping *qWCR4*, 14 recombinants between RM241 and RM255 were firstly screened from a BC_5F_2 population which derived from self-pollinated a BC_5F_1 plant with a heterozygous *qWCR4* segment. After the *qWCR4* interval was narrowed to between markers M24 and M25, a total of 25 recombinants in the M24–M25 region were screened from a BC_5F_3 population which derived from a self-pollinated BC_5F_2 plant with a heterozygous *qWCR4* segment. All recombinant progeny lines of BC_5F_3 and BC_5F_4 generations were used to compare the WCR effect between homologous with J23B and BL130 alleles and each recombinant progeny line was planted in 8 rows with 12 plants per row.

The BC_5F_2, BC_5F_3, and BC_5F_4 populations of *qWCR4* were planted under natural field conditions at the experimental station of Huazhong Agricultural University in Wuhan, Hubei Province in 2014, 2016, and 2018, respectively. The 30-days-old progeny seedlings of each recombinant were transplanted into 8 rows and 12 seedlings each row. The transplanting distance between individual plants of single-row in the field was 16.5 cm, and the row spacing was 26.4 cm. Field management followed local practices. Ten plants were harvested in the middle of each row for traits measurement.

2.2. Phenotyping

Plant height was measured in the field, and harvested panicles from each plant were air-dried and stored at room temperature for 3 months before further phenotyping of other traits. Effective tiller number, panicle length, primary branches, secondary branches, spikelet number per plant, and filled grain number per plant were measured for each plant before threshing. The setting rate was calculated by filled grains per plant divided by spikelet number per plant. Grain length, grain width, thousand-grain weight, and yield per plant were measured after threshing. A total of 60 plants in each near-isogenic line were used for the phenotyping of these traits.

More than 100 seeds from each normal mature plant were randomly selected to be processed into milled rice, and the seed white-core rate of milled rice was investigated through visual observation. Milled rice powder was used to determine the contents of amylose and four kinds of storage proteins (albumin, globulin, prolamin, and glutelin). The method for determining amylose content and 4 storage protein contents was mentioned previously [29,30].

2.3. Genotyping and Sequence Analysis

The parent varieties BL130 and J23B were sequenced by the illumine HiSeq2000 (Illumina, San Diego, CA, USA), and the sequencing data were compared and assembled according to the rice reference genome (Rice Genome Annotation Project, http://rice.uga. edu/, accessed on 6 February 2022) [31]. All mapping primers were designed by primer premier (version 6.0, PREMIER Biosoft, San Francisco, CA, USA) software with reference to the sequencing data of two parents. According to the cetyltrimethylammonium bromide (CTAB) method, genomic DNA was extracted from leaves [32]. The initial program of PCR amplification was 94 °C for 5 min, then 32 cycles of 30 s at 94 °C, 30 s at 55 °C, 30 s at 72 °C, and finally 5 min at 72 °C. The PCR products were identified by sequencing or 4% non-denaturing polyacrylamide gel electrophoresis (PAGE). DNA bands on PAGE gel were displayed by silver nitrate staining and NaOH-formaldehyde solution. Polymorphic simple sequence repeat (SSR) or insertion/deletion (Indel) markers in the QTL interval were used to identify individual genotypes. Relevant primer information can be found in Table S1.

High-quality sequencing bam documents of parents were opened and analyzed in Integrative Genomics Viewer (IGV) software (Broad Institute, Cambridge, MA, USA) [33], and all variations of candidate genes were confirmed by sequencing. Primers used to identify sequence consistency of parents and NILs were listed in Table S1, and the corresponding results were shown in Figure S3.

2.4. RNA Extraction, Reverse Transcription, and qRT-PCR

Expression levels of genes in the M24-M25 interval were examined between NILs of *qWCR4*, which were derived from two homologous progeny of SH60 in the BC_5F_4 population. RNA extraction kit (TRIzol, Invitrogen, Carlsbad, CA, USA) was used for total RNA extraction from different endosperm stages. The first strand of cDNA was synthesized in 20 μL M-MLV reverse transcriptase reaction system (containing 2 μg RNA and 200 U M-MLV reverse transcriptase (Promega, Madison, WI, USA)). Real-time PCR was carried out using Bio-Rad T100TM real-time PCR system (Bio-Rad, Hercules, CA, USA) with the SYBR Green I mix (TaKaRa, Shiga, Japan) on the QuantStudio6Flex instrument (Applied

Biosystems, Carlsbad, CA, USA). All biological tests were repeated at least three times, with three technical repeats, and the rice *Actin1* gene was used as the internal reference. Relevant primers of this analysis are given in Table S2.

2.5. Scanning Electron Microscopy (SEM)

The rice grains of two near-isogenic lines (NILs) were cut in the middle section and gilded under the vacuum condition. The morphology of starch granules in the center of NILs' endosperm was observed by a scanning electron microscope (JSM-6390LV, Jeol Ltd., Tokyo, Japan) under 10 kV acceleration voltage and 30 nm spot size. Scanning electron microscope analyses were performed with at least three biological repeats. All procedures were carried out in accordance with the manufacturer's instructions.

2.6. Statistical Analysis

In the NIL-F_2 population of *qWCR4*, the linkage map of the RM241–RM255 interval was constructed using Mapmaker/Exp3.0 software with the Kosambi mapping function [34], and the *qWCR4* effect detection on WCR was performed by composite interval mapping (CIM) method using Windows QTL cartographer 2.5 software [35]. In terms of fine mapping population, we performed progeny tests to evaluate the WCR effect in recombinant progeny lines. The WCR difference between plants with two different homozygous genotypes was compared by *t*-test. If the *p*-value was less than 0.05, the candidate gene was considered to be located in the heterozygous fragment, otherwise it was considered to be located in the homozygous fragment. Differences in agronomy traits of NILs were analyzed by *t*-test.

2.7. Evolution Analisis

The evolutionary history was inferred using the Neighbor-Joining method [36]. The optimal tree is shown. The tree is drawn to scale, with branch lengths in the same units as the evolutionary distances used to infer the phylogenetic tree. The evolutionary distances were computed using the Poisson correction method [37] and are in the units of the number of amino acid substitutions per site. The proportion of sites where at least 1 unambiguous base is present in at least 1 sequence for each descendent clade is shown next to each internal node in the tree. This analysis involved 13 amino acid sequences. All ambiguous positions were removed for each sequence pair (pairwise deletion option). There was a total of 830 positions in the final dataset. Evolutionary analyses were conducted in MEGA11 [38].

3. Results

3.1. Genetic Validation of qWCR4

In a previous study, *qWCR4* was located between markers RM241 and RM255 on chromosome 4 using a RIL population derived from a cross between BL130 and J23B [17]. To validate the genetic effect of *qWCR4*, a BC_5F_1 plant that carried a heterozygous *qWCR4* region was self-pollinated to obtain the BC_5F_2 population defined as the NIL-F_2 population. Compared with BL130, J23B showed less white belly rate, higher white-core rate, and smaller grain size (Figure S1). Similarly, NIL with homologous J23B allele (NIL(J23B)) had a significantly higher WCR than that of NIL with homologous BL130 allele (NIL(BL130)) in the NIL-F_2 population of *qWCR4* (Figure 1a). We also investigated the starch granules morphology in the endosperm center between two homologous NILs. As shown in Figure 1b, starch granules in the endosperm center of white-core grains from NIL(J23B) were small, round, and loosely packed, which were notably different from NIL(BL130) with polyhedral and densely packed starch granule morphology (Figure 1b). Additionally, *qWCR4* showed an incompletely dominant effect (Figure 1c), and the phenotypic variation of WCR in the BC_5F_2 population showed continuous bimodal distribution (Figure 1d). *qWCR4* was mapped to the region between RM241 and RM255 and explained 21.1% of the phenotypic variance (Table 1).

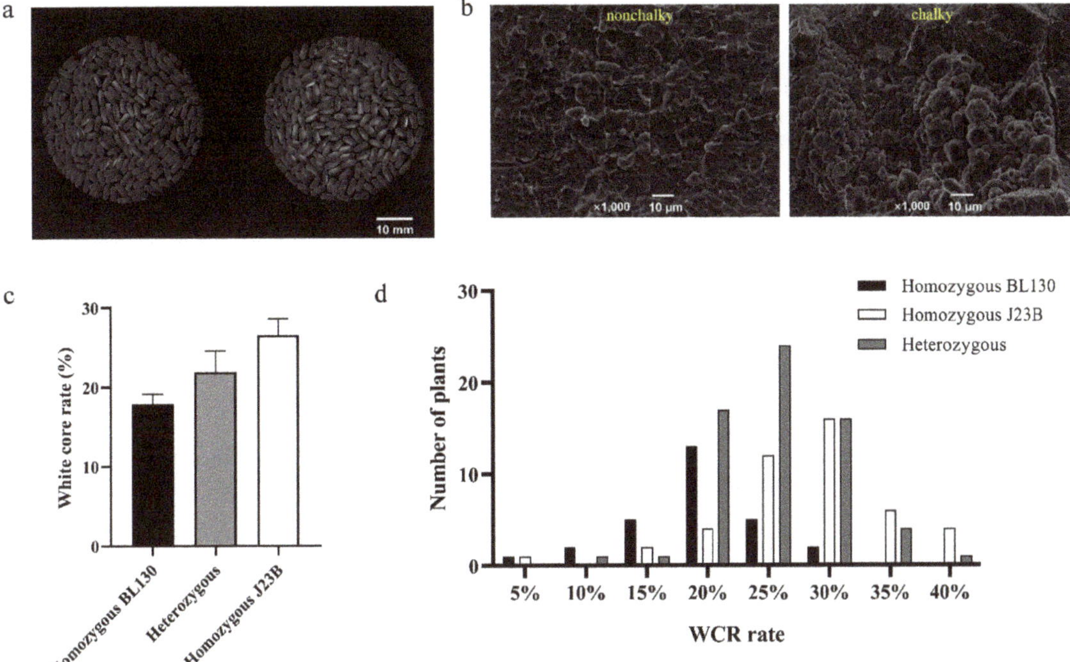

Figure 1. Genetic validation of QTL effects of *qWCR7* in BC₅F₂ population. (**a**) Chalkiness performance of NIL(BL130) (left) and NIL(J23B) (right). (**b**) SEM analysis of nonchalky (left) and chalky (lower right) grains of NIL(BL130) and NIL(J23B). (**c**) WCR difference analysis among three genotypes. (**d**) Distribution of WCR performance variation.

Table 1. Genetic effect of *qWCR4* in BC₅F₂ population.

Trait	QTL	Interval	LOD	A	V (%)
WCR	*qWCR4*	RM241–RM255	8.95	−4.19	21.1

WCR, white core rate. QTL, quantitative trait locus. LOD, logarithms of odds. A, additive effect, the negative value means that J23B allele increases the trait value. V, variance, phenotypic variation explained by the QTL.

These results indicated that *qWCR4* was the genetic factor responsible for WCR variation, and the BL130-derived allele decreased the endosperm WCR.

3.2. Fine Mapping of qWCR4

We made a two-step mapping strategy to improve the efficiency of mapping *qWCR4*. Firstly, 14 recombinants between markers RM241 and RM255 were identified from a BC₅F₂ population consisting of 1000 individuals, and the corresponding progeny population of each recombinant was used for the progeny test. To genotype all recombinants, 7 indel markers in this region were developed based on the sequence variation between two parents (Figure 2a). The progeny test was undertaken to determine the *qWCR4* genotype of each recombinant. According to the progeny testing results of recombinant lines ZY03, ZY20, and ZY27, which have similar crossover intervals between M24 and M25, we narrowed the *qWCR4* to a 536 kb region between M24 and M25 (Figure 2a).

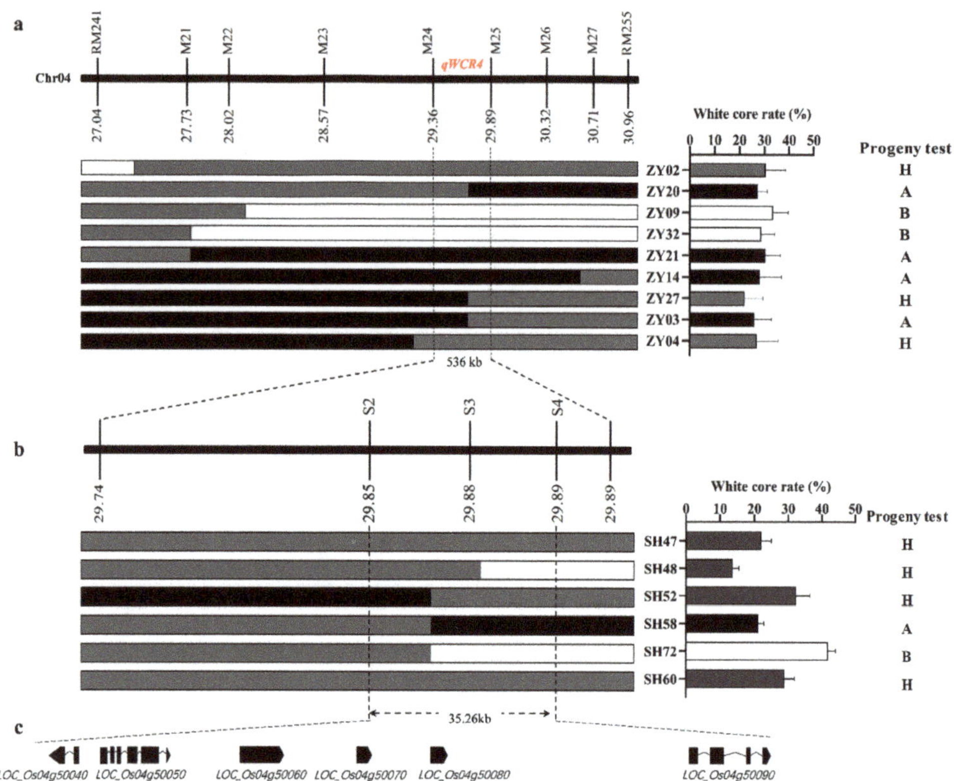

Figure 2. Fine mapping of *qWCR4*. (**a**) The first-step fine mapping of *qWCR4*. (**b**) The second-step fine mapping of *qWCR4*. (**c**) Schematic representation of candidate gene in *qWCR4* region. Black, white, and grey blocks represent the genotypes of homozygous BL130, homozygous J23B, and heterozygote, respectively. Genotypes and phenotypes of recombinants, each of which was confirmed by progeny test, the A and B of the progeny test meant no WCR difference between two homologous plants in recombinant progeny, while the H of the progeny test meant significant WCR difference.

After the 536 kb region was obtained from the first step, a BC_5F_3 population consisting of 2500 individuals was used to screen new recombinants for the second-step fine mapping. A total of 25 recombinants between M24 and M25 were identified in this population, and 3 indel markers in this region were developed and used to genotype these recombinants. As shown in Figure 2b, progeny testing results of 4 recombinants that crossover between markers S2 and S4 indicated that the candidate gene underlying for WCR in the *qWCR4* region should be in the 35.26 kb region between these two markers.

3.3. Expression and Sequence Analysis of Candidate Genes

According to the annotation information of the *japonica* variety Nipponbare in RGAP website (Rice Genome Annotation Project, http://rice.uga.edu/, accessed on 6 February 2022), the target 35.26 kb region of *qWCR4* contains six annotated genes (*LOC_Os04g50040*, *LOC_Os04g50050*, *LOC_Os04g50060*, *LOC_Os04g50070*, *LOC_Os04g50080*, and *LOC_Os04g50090*) (Figure 2c). Among these genes, *LOC_Os04g50060*, *LOC_Os04g50070*, and *LOC_Os04g50090* encode a GRAS family transcription factor domain-containing protein, a ZOS4-13-C2H2 zinc finger protein, and a putative helix-loop-helix DNA-binding protein, respectively. The remaining genes encode two expressed proteins (*LOC_Os04g50040* and *LOC_Os04g50080*) and one putative retrotransposon protein (*LOC_Os04g50050*). According to the RNA-Seq data in the RGAP website, the two expressed proteins had no expression in endosperm or seed, and retrotransposon was usually not considered a functional gene. Therefore, we paid attention to three functional genes expressed in the endosperm.

We then examined the relative expression level of these three genes in different stage endosperm by quantitative real-time PCR (Figure 3a). The result showed that *LOC_Os04g50090* was not expressed in the endosperm. *LOC_Os04g50060* was notably expressed at a relatively high level throughout the whole endosperm development period with higher expression in NIL(BL130). *LOC_Os04g50070* expressed only in 5E (endosperm 5 days after pollination) with a significant difference between NILs. We also analyzed the nucleotide diversity of two parents in *LOC_Os04g50060* and *LOC_Os04g50070*. The genomic region used for nucleotide diversity analysis contains the promotor (~2 kb) and the entire ORF of two genes. According to the sequencing data of two parents, 22 SNP variants were found in the *LOC_Os04g50060* region, of which 17 were in the promotor and 5 in the ORF (Figure 3b). 9 SNP variants were found in the *LOC_Os04g50070* region, of which 8 were in the promotor and 1 in the ORF. Evolutionary analysis showed that *LOC_Os04g50060* and *LOC_Os04g50070* exhibited high homology with GRAS family transcription factor and zinc finger protein family, respectively (Figure S4). Taken together, both *LOC_Os04g50060* and *LOC_Os04g50070* could be the candidate genes for WCR in the *qWCR4* region.

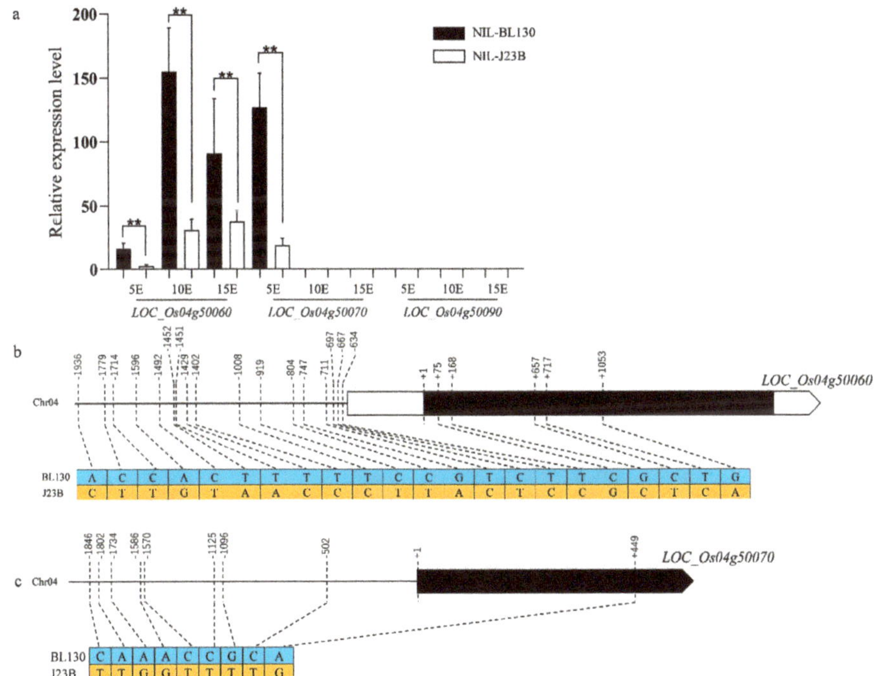

Figure 3. Expression and sequence variations of two candidate genes in *qWCR4* locus. (**a**) Relative expression level analysis of three candidate genes. *LOC_Os04g50060*, *LOC_Os04g50070*, and *LOC_Os04g50090* represent three causal genes underlying the *qWCR4* locus. 5E, 10E, and 15E mean endosperm 5 days, 10 days, and 15 days after pollination, respectively. ** means significant differences at $p < 0.01$, student's *t*-tests. (**b**) Variation analysis of *LOC_Os04g50060*. (**c**) Variation analysis of *LOC_Os04g50070*. Boxes filled with black and white in (**b**) and (**c**) mean coding and untranslated regions, respectively. The solid lines on the left side of boxes in (**b**) and (**c**) mean the promotor region. The dotted lines on the solid lines and boxes in (**b**) and (**c**) represent the position from the translation initiation site (ATG), + and − mean located in down-stream and up-stream of ATG.

3.4. Agronomic Traits of Two NILs

The abnormal starch granule morphology may be caused by a developmental disorder. Researchers have proved that high filling speed decreased rice quality [39,40]. Here, we investigated the seed filling speed after pollination of two NILs. NIL(J23B) with higher WCR accompanied with a higher filling speed in the middle and later stage (from 12 days after pollination to 20 days after pollination) (Figure 4a). This result implied that quicker endosperm filling might result in abnormal starch granule morphology.

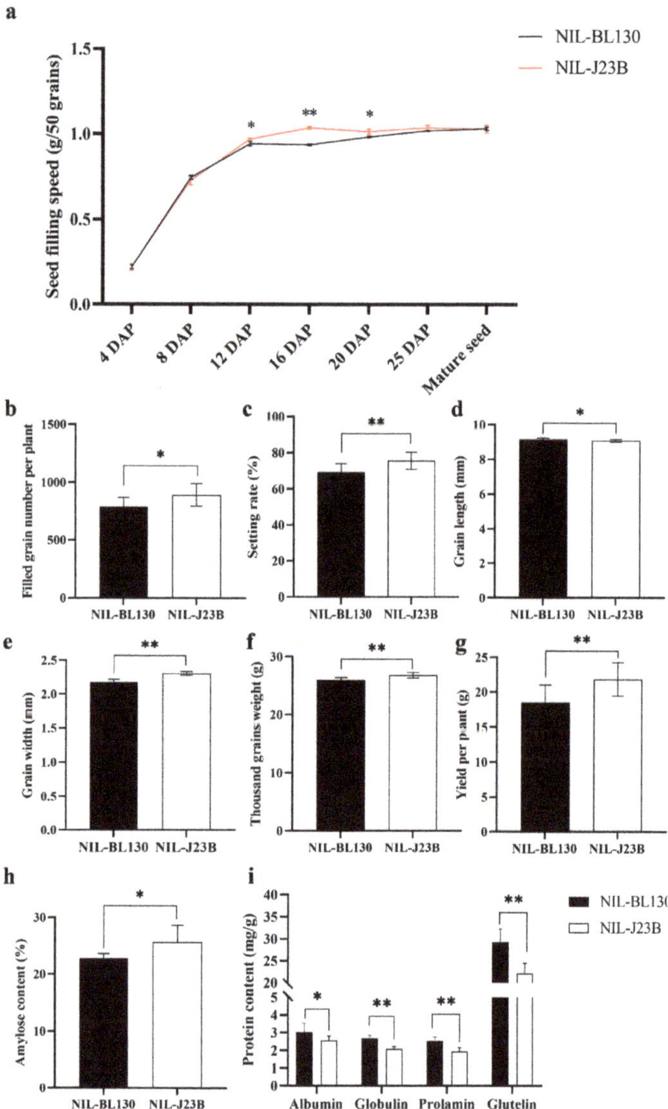

Figure 4. Trait performance of two NILs. Differences in Seed filling rate (**a**), Filling grain number per plant (**b**), Setting rate (**c**), Grain length (**d**), Grain width (**e**), Thousand grain weight (**f**), Yield per plant (**g**), Amylose content (**h**) and protein content (**i**) were analysised between two NILs. * and ** mean significant differences at $p < 0.05$ and $p < 0.01$ respectively, student's t-tests.

Chalkiness is closely related to grain size and quality traits, such as grain length, grain width, starch, and protein content [19,26,27]. In the present study, we investigated agronomy traits of NILs to detect the effect of *qWCR4* on these traits. No significant differences were observed in plant height, effective tiller number, panicle length, primary branches number, secondary branches number, and spikelet number per plant, leading to similar plant architecture between these two NILs (Figure S2a–f). However, compared with the NIL(J23B), NIL(BL130) with lower WCR had more grain length but less filled grain number, setting rate, and grain width, leading to an obviously decreased thousand-grain weight and yield per plant (Figure 4b–g). In addition, we also examined the amylose content and four storage protein content of these two NILs. NIL(BL130) showed decreased amylose content but higher storage protein content (Figure 4h,i). The above results suggested that the *qWCR4* allele from J23B has a positive effect on yield-related traits and storage content but a negative effect on amylose content and chalkiness.

4. Discussion

Unlike waxy grain, chalky grain is attributed to the air gap between irregular starch granules, which cause scatter light, leading to an opaque phenomenon [41]. However, the opacity of waxy rice is due to the diffusion of light from micropores and hollows in starch granules [42]. The external performance and internal characteristics of rice dramatically influence consumer choice. As one of the evaluation indexes of poor-quality rice, chalkiness not only affects the appearance of rice [43–45] but also greatly affects the rice milling yield [4,5,8] and cooking and eating quality [6]. Therefore, understanding the formation mechanism of endosperm chalkiness is essential to rice yield and quality.

4.1. A New QTL Was Found to Control WCR in Rice Endosperm

Many chalkiness QTLs were identified in the past decades, but only a few QTLs/genes related to chalkiness have been fine mapped/cloned. There are two possible reasons. Firstly, chalkiness is a quantitative trait controlled by polygenes, abnormal seed filling, sugar transport, and starch synthase could lead to the chalky phenomenon. Second, the chalkiness is seriously affected by an especially high temperature environment. Therefore, researchers usually used different populations under different environments to obtain stable chalkiness QTLs [11,12,14,17,46]. In a previous study, *qWCR4* was repeatedly detected in two environments and confirmed in a NIL-F$_2$ population [17]. In this study, *qWCR4* was confirmed in a higher generation genetic population (Figure 1) and showed a higher phenotypic variation explained by the QTL (Table 1). Starch granules of chalky grains from NIL(J23B) exhibited a typically chalky morphology according to previous research (Figure 1b). Through two-step fine mapping, *qWCR4* was narrowed to a 35.26 kb region between markers S2 and S4 (Figure 2a,b). Until now, few chalkiness-related genes have been reported on chromosome 4, while *qWCR4* is a newly found QTL affecting chalkiness in this region. Therefore, we found a new QTL *qWCR4* controlling WCR in chromosome 4.

4.2. LOC_Os04g50060 and LOC_Os04g50070 Could Be the Candidate Genes for qWCR4

According to the annotation data on the RGAP website, six genes exist in the *qWCR4* region. Two of them encoded expressed protein not expressed in the endosperm. *LOC_Os04g50050* encoded a putative transposon not considered a functional gene in general. *LOC_Os04g50090* encoded an HLH DNA binding protein not expressed in endosperm. *LOC_Os04g50060* encoded a GRAS family transcription factor with a dramatic expression change in different endosperm stages, and showed high protein homology with identified GRAS family transcription factors in other species. *LOC_Os04g50070* encoding a C2H2 zinc finger protein with significant expression changed only in 5E, and also exhibited protein homology with a reported zinc finger protein family. Cai et al. [47] reported a GRAS family transcription factor *ZmGRAS20*, which is expressed in endosperm specifically, may function as a starch synthesis regulatory factor in rice endosperm. *ZmGRAS20* transgenic seeds exhibited altered starch granules morphology, and overexpression of *ZmGRAS20* led to a chalky region of ventral endosperm with decreased starch content. Ji et al. [48] discovered an *Opaque 2 (O2)-ZmGRAS11-ZmEXPB12* regulatory module that regulates endosperm cell expansion and endosperm filling. On the other hand, Royo et al. [49] identified two closely related C2H2-type zinc finger proteins, *ZmMRPI-1* and *ZmMRPI-2*, which interact with *ZmMRP-1* and modulate its activity on transfer cell-specific promotors. Jiang et al. [50] found *ZmZAT8*, a conserved feature of plant C2H2-type zinc finger protein, plays a positive role in regulating starch synthesis in maize endosperm and could be strongly stimulated by ABA. Taken together, both GRAS transcription factor and C2H2-type zinc finger protein could be the candidate gene control WCR in *qWCR4* interval. Sequence variation analysis between two parents showed that 22 and 9 SNPs exist in *LOC_Os04g50060* and *LOC_Os04g50070*, respectively. Variations in the promotor of these two genes may lead to differences in the expression levels, and ultimately affects the WCR performance of two NILs.

4.3. qWCR4 Changed the Yield and Quality Performance of NILs

Many studies have shown that high filling speed would decrease the quality of rice and increase the rice chalky rate [39,40]. Compared to NIL(BL130), NIL(J23B) showed higher seed filling speed (Figure 4a), and then more white-core grains occurred. NIL with higher WCR showed superiority in filled grain number, grain width, and TGW, which resulted in a significant advantage in yield per plant (Figure 4b–g). According to previous reports, the GRAS family transcription factor and C2H2-type zinc finger protein may also have a function in regulating grain width [51,52], and many studies have shown that the increase of grain width increases the WCR [18,19]. Therefore, whether the higher WCR in NIL(J23B) is caused by the larger grain width needs to be further studied.

Many researchers have reported that rice endosperm with higher WCR is often accompanied by lower starch [9] and protein content [4]. We further analyzed the content of amylose and storage proteins in NILs, and higher WCR was found along with higher amylose content and lower storage protein content (Figure 4h,i). It may be because *qWCR4* affects the amylose content and storage protein content through other mechanisms. The increase of grain width is accompanied by increased amylose content [53]. *qWCR4* may affect amylose and storage protein content by increasing grain width. All in all, fine-mapping *qWCR4* could provide gene resources for breeding high-yield and quality rice varieties.

5. Conclusions

Our research revealed that *qWCR4* was a genetic factor conferring WCR variation and was narrowed to a 35.26 kb region. The quantitative RT-PCR and sequence variation analysis of genes in the *qWCR4* region showed that *LOaC_Os04g50060* and *LOC_Os04g50070*, coding a GRAS and a C2H2 family transcription factor in the *qWCR4* region, respectively, may be candidate genes for *qWCR4*. Different *qWCR4* alleles significantly changed the agronomic traits of NILs. *qWCR4^{J23B}* could significantly increase the seed setting rate and grain width of NIL(J23B), resulting in a higher yield per plant, but also an increase in WCR of about 9%. However, whether more chalky grains are caused by higher seed filling speed or wider grain needs further study. Finally, the results of this study also laid a foundation for map-based cloning of *qWCR4*.

Supplementary Materials: The following supporting information can be downloaded at: https://www.mdpi.com/article/10.3390/agronomy12030706/s1, Table S1: Primers used for fine mapping and NILs identification; Table S2: Primers used for quantitative RT-PCR; Figure S1: Grain characters of parents and NILs; Figure S2: Agronomic traits difference of NILs; Figure S3: Sequence consistency of NILs and parents; Figure S4: Evolutionary tree of *LOC_Os04g50060* with GRAS family transcription factor (a) and *LOC_Os04g50070* with zinc-finger protein family (b).

Author Contributions: Conceptualization, H.S., Y.H., Y.Z. and P.Y.; methodology, H.S., Y.H., Y.Z. and P.Y.; investigation, H.S., Y.Z., L.W. and Y.W.; resources, Y.H., P.Y., G.G., Q.Z. and X.L.; writing-original draft preparation, H.S.; writing-review and editing, H.S., Y.H. and G.L.; supervision, Y.H. All authors have read and agreed to the published version of the manuscript.

Funding: This work was supported by grants from the National Natural Science Foundation of China (U21A20211), the Ministry of Science and Technology (2021YFF1000200, 2020YFD0900302), the science and technology major program of Hubei Province (2021ABA011, 2020BBB051), and the China Agriculture Research System (CARS-01-03).

Institutional Review Board Statement: Not applicable.

Informed Consent Statement: Not applicable.

Data Availability Statement: The datasets generated during the current study are available from the corresponding author on reasonable request.

Acknowledgments: The authors extend their appreciation for the support from the National Key Laboratory of Crop Genetic Improvement, Huazhong Agricultural University.

Conflicts of Interest: Authors declare that there are no conflict of interest.

References

1. Xu, L.; Yuan, S.; Man, J. Changes in rice yield and yield stability in China during the past six decades. *J. Sci. Food Agric.* **2020**, *100*, 3560–3569. [CrossRef] [PubMed]
2. Gao, Y.; Liu, C.L.; Li, Y.Y.; Zhang, A.P.; Dong, G.J.; Xie, L.H.; Zhang, B.; Ruan, B.P.; Hong, K.; Xue, D.W.; et al. QTL analysis for chalkiness of rice and fine mapping of a candidate gene for *qACE9*. *Rice* **2016**, *9*, 41. [CrossRef]
3. Yoshioka, Y.; Iwata, H.; Tabata, M.; Ninomiya, S.; Ohsawa, R. Chalkiness in Rice: Potential for Evaluation with Image Analysis. *Crop Sci.* **2007**, *47*, 2113–2120. [CrossRef]
4. Lanning, S.B.; Siebenmorgen, T.J.; Counce, P.A.; Ambardekar, A.A.; Mauromoustakos, A. Extreme nighttime air temperatures in 2010 impact rice chalkiness and milling quality. *Field Crops Res.* **2011**, *124*, 132–136. [CrossRef]
5. Butardo, V.M.; Sreenivasulu, N. Improving Head Rice Yield and Milling Quality: State-of-the-Art and Future Prospects. In *Rice Grain Quality: Methods and Protocols*; Sreenivasulu, N., Ed.; Springer: New York, NY, USA, 2019; pp. 1–18.
6. Fitzgerald, M.A.; Resurreccion, A.P. Maintaining the yield of edible rice in a warming world. *Funct. Plant Biol.* **2009**, *36*, 1037–1045. [CrossRef] [PubMed]
7. Zhang, S.S.; Zhan, J.P.; Yadegari, R. Maize opaque mutants are no longer so opaque. *Plant Reprod.* **2018**, *31*, 319–326. [CrossRef] [PubMed]
8. Yamakawa, H.; Hirose, T.; Kuroda, M.; Yamaguchi, T. Comprehensive expression profiling of rice grain filling-related genes under high temperature using DNA microarray. *Plant Physiol.* **2007**, *144*, 258–277. [CrossRef] [PubMed]
9. Chun, A.; Song, J.; Kim, K.J.; Lee, H.J. Quality of head and chalky rice and deterioration of eating quality by chalky rice. *J. Crop Sci. Biotechnol.* **2009**, *12*, 239–244. [CrossRef]
10. Sreenivasulu, N.; Butardo, V.M., Jr.; Misra, G.; Cuevas, R.P.; Anacleto, R.; Kavi Kishor, P.B. Designing climate-resilient rice with ideal grain quality suited for high-temperature stress. *J. Exp. Bot.* **2015**, *66*, 1737–1748. [CrossRef] [PubMed]
11. He, P.; Li, S.G.; Qian, Q.; Ma, Y.Q.; Li, J.Z.; Wang, W.M.; Chen, Y.; Zhu, L.H. Genetic analysis of rice grain quality. *Theor. Appl. Genet.* **1999**, *98*, 502–508. [CrossRef]
12. Chen, H.M.; Zhao, Z.G.; Jiang, L.; Wan, X.Y.; Liu, L.L.; Wu, X.J.; Wan, J.M. Molecular genetic analysis on percentage of grains with chalkiness in rice (*Oryza sativa* L.). *Afr. J. Biotechnol.* **2011**, *10*, 6891–6903. [CrossRef]
13. Chen, Y.; Wang, M.; Ouwerkerk, P.B. Molecular and environmental factors determining grain quality in rice. *Food Energy Secur.* **2012**, *1*, 111–132. [CrossRef]
14. Zhao, X.Q.; Daygon, V.D.; McNally, K.L.; Hamilton, R.S.; Xie, F.M.; Reinke, R.F.; Fitzgerald, M.A. Identification of stable QTLs causing chalk in rice grains in nine environments. *Theor. Appl. Genet.* **2016**, *129*, 141–153. [CrossRef] [PubMed]
15. Zhu, A.K.; Zhang, Y.X.; Zhang, Z.H.; Wang, B.F.; Xue, P.; Cao, Y.R.; Chen, Y.Y.; Li, Z.H.; Liu, Q.E.; Cheng, S.H. Genetic dissection of *qPCG1* for a quantitative trait locus for percentage of chalky grain in rice (*Oryza sativa* L.). *Front. Plant Sci.* **2018**, *9*, 1173. [CrossRef] [PubMed]
16. Misra, G.; Badoni, S.; Parween, S.; Singh, R.K.; Leung, H.; Ladejobi, O.; Mott, R.; Sreenivasulu, N. Genome-wide association coupled gene to gene interaction studies unveil novel epistatic targets among major effect loci impacting rice grain chalkiness. *Plant Biotechnol. J.* **2021**, *19*, 910–925. [CrossRef] [PubMed]
17. Yun, P.; Zhu, Y.; Wu, B.; Gao, G.J.; Sun, P.; Zhang, Q.J.; He, Y.Q. Genetic mapping and confirmation of quantitative trait loci for grain chalkiness in rice. *Mol. Breed.* **2016**, *36*, 1–8. [CrossRef]
18. Okada, S.; Iijima, K.; Hori, K.; Yamasaki, M. Genetic and epistatic effects for grain quality and yield of three grain-size QTLs identified in brewing rice (*Oryza sativa* L.). *Mol. Breed.* **2020**, *40*, 1–12. [CrossRef]
19. Wang, Y.X.; Xiong, G.S.; Hu, J.; Jiang, L.; Yu, H.; Xu, J.; Fang, Y.X.; Zeng, L.J.; Xu, E.B.; Xu, J. Copy number variation at the *GL7* locus contributes to grain size diversity in rice. *Nat. Genet.* **2015**, *47*, 944–948. [CrossRef]
20. Tanaka, N.; Fujita, N.; Nishi, A.; Satoh, H.; Hosaka, Y.; Ugaki, M.; Kawasaki, S.; Nakamura, Y. The structure of starch can be manipulated by changing the expression levels of starch branching enzyme IIb in rice endosperm. *Plant Biotechnol. J.* **2004**, *2*, 507–516. [CrossRef]

21. Wang, Y.H.; Ren, Y.L.; Liu, X.; Jiang, L.; Chen, L.M.; Han, X.H.; Jin, M.N.; Liu, S.J.; Liu, F.; Lv, J.; et al. *OsRab5a* regulates endomembrane organization and storage protein trafficking in rice endosperm cells. *Plant J.* **2010**, *64*, 812–824. [CrossRef] [PubMed]

22. Liu, F.; Ren, Y.L.; Wang, Y.H.; Peng, C.; Zhou, K.N.; Lv, J.; Guo, X.P.; Zhang, X.; Zhong, M.S.; Zhao, S.L. *OsVPS9A* functions cooperatively with *OsRAB5A* to regulate post-Golgi dense vesicle-mediated storage protein trafficking to the protein storage vacuole in rice endosperm cells. *Mol. Plant* **2013**, *6*, 1918–1932. [CrossRef] [PubMed]

23. Ren, Y.L.; Wang, Y.H.; Liu, F.; Zhou, K.N.; Ding, Y.; Zhou, F.; Wang, Y.; Liu, K.; Gan, L.; Ma, W.W. *GLUTELIN PRECURSOR ACCUMULATION3* encodes a regulator of post-Golgi vesicular traffic essential for vacuolar protein sorting in rice endosperm. *Plant Cell* **2014**, *26*, 410–425. [CrossRef]

24. Wei, X.J.; Jiao, G.A.; Lin, H.Y.; Sheng, Z.H.; Shao, G.N.; Xie, L.H.; Tang, S.Q.; Xu, Q.G.; Hu, P.S. *GRAIN INCOMPLETE FILLING 2* regulates grain filling and starch synthesis during rice caryopsis development. *J Integr. Plant Biol.* **2017**, *59*, 134–153. [CrossRef] [PubMed]

25. Lou, G.M.; Chen, P.L.; Zhou, H.; Li, P.B.; Xiong, J.W.; Wan, S.S.; Zheng, Y.Y.; Alam, M.; Liu, R.J.; Zhou, Y.; et al. *FLOURY ENDOSPERM19* encoding a class I glutamine amidotransferase affects grain quality in rice. *Mol. Breed.* **2021**, *41*, 36. [CrossRef]

26. Li, Y.B.; Fan, C.C.; Xing, Y.Z.; Yun, P.; Luo, L.J.; Yan, B.; Peng, B.; Xie, W.B.; Wang, G.W.; Li, X.H.; et al. *Chalk5* encodes a vacuolar H^+-translocating pyrophosphatase influencing grain chalkiness in rice. *Nat. Genet.* **2014**, *46*, 398–404. [CrossRef]

27. Wu, B.; Xia, D.; Zhou, H.; Cheng, S.Y.; Wang, Y.P.; Li, M.Q.; Gao, G.J.; Zhang, Q.L.; Li, X.H.; He, Y.Q. Fine mapping of *qWCR7*, a grain chalkiness QTL in rice. *Mol. Breed.* **2021**, *41*, 1–12. [CrossRef]

28. Wu, B.; Yun, P.; Zhou, H.; Xia, D.; Gu, Y.; Li, P.B.; Yao, J.L.; Zhou, Z.Q.; Chen, J.X.; Liu, R.J.; et al. Natural variation in *WHITE-CORE RATE 1* regulates redox homeostasis in rice endosperm to affect grain quality. *Plant Cell* **2022**, koac057. [CrossRef] [PubMed]

29. Kumamaru, T. Mutants for rice storage proteins 1. Screening of mutants semidwarfism-related proteins and glutelin seed protein in rice (*Oryza sativa* L.). *Theor. Appl. Genet.* **1988**, *83*, 153–158. [CrossRef]

30. Bao, J.; Shen, S.; Sun, M.; Corke, H. Analysis of genotypic diversity in the starch physicochemical properties of nonwaxy rice: Apparent amylose content, pasting viscosity and gel texture. *Starch* **2006**, *58*, 259–267. [CrossRef]

31. Li, H.; Durbin, R. Fast and accurate short read alignment with Burrows–Wheeler transform. *Bioinformatics* **2009**, *25*, 1754–1760. [CrossRef]

32. Murray, M.G.; Thompson, W.F. Rapid isolation of high molecular weight plant DNA. *Nucleic Acids Res.* **1980**, *8*, 4321–4326. [CrossRef]

33. Robinson, J.T.; Thorvaldsdóttir, H.; Winckler, W.; Guttman, M.; Lander, E.S.; Getz, G.; Mesirov, J.P. Integrative genomics viewer. *Nat. Biotechnol.* **2011**, *29*, 24–26. [CrossRef] [PubMed]

34. Lander, E.S.; Green, P.; Abrahamson, J.; Barlow, A.; Daly, M.J.; Lincoln, S.E.; Newburg, L. MAPMAKER: An interactive computer package for constructing primary genetic linkage maps of experimental and natural populations. *Genomics* **1987**, *1*, 174–181. [CrossRef]

35. Zeng, Z.B. Precision mapping of quantitative trait loci. *Genetics* **1994**, *136*, 1457–1468. [CrossRef]

36. Saitou, N.; Nei, M. The neighbor-joining method: A new method for reconstructing phylogenetic trees. *Mol. Biol. Evol.* **1987**, *4*, 406–425. [CrossRef] [PubMed]

37. Zuckerkandl, E.; Pauling, L. Evolutionary divergence and convergence in proteins. In *Evolving Genes and Proteins*; Elsevier: Amsterdam, The Netherlands, 1965; pp. 97–166.

38. Tamura, K.; Stecher, G.; Kumar, S. MEGA11: Molecular evolutionary genetics analysis version 11. *Mol. Biol. Evol.* **2021**, *38*, 3022–3027. [CrossRef]

39. Ito, S.; Hara, T.; Kawanami, Y.; Watanabe, T.; Thiraporn, K.; Ohtake, N.; Sueyoshi, K.; Mitsui, T.; Fukuyama, T.; Takahashi, Y. Carbon and nitrogen transport during grain filling in rice under high-temperature conditions. *J. Agron. Crop Sci.* **2009**, *195*, 368–376. [CrossRef]

40. Chen, J.L.; Tang, L.; Shi, P.H.; Yang, B.H.; Sun, T.; Cao, W.X.; Zhu, Y. Effects of short-term high temperature on grain quality and starch granules of rice (*Oryza sativa* L.) at post-anthesis stage. *Protoplasma* **2017**, *254*, 935–943. [CrossRef]

41. Custodio, M.C.; Cuevas, R.P.; Ynion, J.; Laborte, A.G.; Velasco, M.L.; Demont, M. Rice quality: How is it defined by consumers, industry, food scientists, and geneticists? *Trends Food Sci. Technol.* **2019**, *92*, 122–137. [CrossRef] [PubMed]

42. Ashida, K.; Iida, S.; Yasui, T. Morphological, physical, and chemical properties of grain and flour from chalky rice mutants. *Cereal Chem.* **2009**, *86*, 225–231. [CrossRef]

43. Yang, W.; Wu, K.; Wang, B.; Liu, H.; Guo, S.; Guo, X.; Luo, W.; Sun, S.; Ouyang, Y.; Fu, X. The RING E3 ligase CLG1 targets GS3 for degradation via the endosome pathway to determine grain size in rice. *Mol. Plant* **2021**, *14*, 1699–1713. [CrossRef] [PubMed]

44. Asako, K.; Bao, G.L.; Ye, S.H.; Katsura, T. Detection of Quantitative Trait Loci for White-back and Basal-white Kernels under High Temperature Stress in *japonica* Rice Varieties. *Breed. Sci.* **2007**, *57*, 107–116. [CrossRef]

45. Tabata, M.; Hirabayashi, H.; Takeuchi, Y.; Ando, I.; Iida, Y.; Ohsawa, R. Mapping of quantitative trait loci for the occurrence of white-back kernels associated with high temperatures during the ripening period of rice (*Oryza sativa* L.). *Breed. Sci.* **2007**, *57*, 47–52. [CrossRef]

46. Peng, B.; Wang, L.Q.; Fan, C.C.; Jiang, G.H.; Luo, L.J.; Li, Y.B.; He, Y.Q. Comparative mapping of chalkiness components in rice using five populations across two environments. *BMC Genet.* **2014**, *15*, 1–14. [CrossRef]

47. Cai, H.L.; Chen, Y.L.; Zhang, M.; Cai, R.H.; Cheng, B.J.; Ma, Q.; Zhao, Y. A novel *GRAS* transcription factor, *ZmGRAS20*, regulates starch biosynthesis in rice endosperm. *Physiol. Mol. Biol. Plants* **2017**, *23*, 143–154. [CrossRef]
48. Ji, C.; Xu, L.N.; Li, Y.J.; Fu, Y.X.; Li, S.; Wang, Q.; Zeng, X.; Zhang, Z.Q.; Zhang, Z.Y.; Wang, W.Q. The *O2-ZmGRAS11* transcriptional regulatory network orchestrates the coordination of endosperm cell expansion and grain filling in maize. *Mol. Plant* **2021**. [CrossRef]
49. Royo, J.; Gómez, E.; Barrero, C.; Muñiz, L.M.; Sanz, Y.; Hueros, G. Transcriptional activation of the maize endosperm transfer cell-specific gene *BETL1* by *ZmMRP-1* is enhanced by two C2H2 zinc finger-containing proteins. *Planta* **2009**, *230*, 807–818. [CrossRef]
50. Jiang, T.L.; Xia, M.; Huang, H.H.; Xiao, J.L.; Long, J.; Li, X.; Zhang, J.J. Identification of transcription factor *zmzat8* involved in abscisic acid regulation pathway of starch synthesis in maize endosperm. *Pak. J. Bot* **2019**, *51*, 2121–2128. [CrossRef]
51. Sun, L.J.; Li, X.J.; Fu, Y.C.; Zhu, Z.F.; Tan, L.B.; Liu, F.X.; Sun, X.Y.; Sun, X.W.; Sun, C.Q. *GS6*, a member of the GRAS gene family, negatively regulates grain size in rice. *J. Integr. Plant Biol.* **2013**, *55*, 938–949. [CrossRef]
52. Xu, Q.K.; Yu, H.P.; Xia, S.S.; Cui, Y.J.; Yu, X.Q.; Liu, H.; Zeng, D.L.; Hu, J.; Zhang, Q.; Gao, Z.Y. The C$_2$H$_2$ zinc-finger protein *LACKING RUDIMENTARY GLUME 1* regulates spikelet development in rice. *Sci. Bull.* **2020**, *65*, 753–764. [CrossRef]
53. Achary, V.M.M.; Reddy, M.K. CRISPR-Cas9 mediated mutation in GRAIN WIDTH and WEIGHT2 (GW2) locus improves aleurone layer and grain nutritional quality in rice. *Sci. Rep.* **2021**, *11*, 21941. [CrossRef] [PubMed]

Article

Molecular Mapping of a New Brown Planthopper Resistance Gene *Bph43* in Rice (*Oryza sativa* L.)

JangChol Kim [1,2,†], Xin An [1,2,†], Ke Yang [1,2], Si Miao [1,2], Yushi Qin [1,2], Yinxia Hu [1,2], Bo Du [1], Lili Zhu [1], Guangcun He [1] and Rongzhi Chen [1,2,*]

[1] State Key Laboratory of Hybrid Rice, College of Life Sciences, Wuhan University, Wuhan 430072, China; kimjangchol@whu.edu.cn (J.K.); anxin-@whu.edu.cn (X.A.); 2019102040039@whu.edu.cn (K.Y.); 2020202040071@whu.edu.cn (S.M.); qys929560@163.com (Y.Q.); huyinxia163@163.com (Y.H.); bodu@whu.edu.cn (B.D.); zhulili58@sina.com (L.Z.); gche@whu.edu.cn (G.H.)
[2] Wuhan University Shenzhen Research Institute, Shenzhen 518057, China
* Correspondence: rzchen@whu.edu.cn
† These authors contributed equally to this work.

Abstract: Brown planthopper (BPH) has become the most devastating insect pests of rice and a serious threat to rice production. To combat newly occurring virulent BPH populations, it is still urgent to explore more new broad-spectrum BPH resistance genes and integrate them into rice cultivars. In the present study, we explored the genetic basis of BPH resistance in IRGC 8678. We identified and mapped a new resistance gene *Bph43* to a region of ~380 kb on chromosome 11. Genes encoding nucleotide-binding domain leucine-rich repeat-containing (NBS-LRR)-type disease resistance proteins or Leucine Rich Repeat family proteins annotated in this region were predicted as the possible candidates for *Bph43*. Meanwhile, we developed near isogenic lines of *Bph43* (NIL-*Bph43*-9311) in an elite restorer line 9311 background using marker-assisted selection (MAS). The further characterization of NIL-*Bph43*-9311 demonstrated that *Bph13* confers strong antibiosis and antixenosis effects on BPH. Comparative transcriptome analysis revealed that genes related to the defense response and resistance gene-dependent signaling pathway were significantly and uniquely enriched in BPH-infested NIL-*Bph43*-9311. Our work demonstrated that *Bph43* can be deployed as a valuable donor in BPH resistance breeding programs.

Keywords: brown planthopper; gene mapping; near isogenic line; marker-assisted selection; RNA-Seq

Citation: Kim, J.; An, X.; Yang, K.; Miao, S.; Qin, Y.; Hu, Y.; Du, B.; Zhu, L.; He, G.; Chen, R. Molecular Mapping of a New Brown Planthopper Resistance Gene *Bph43* in Rice (*Oryza sativa* L.). *Agronomy* **2022**, *12*, 808. https://doi.org/10.3390/agronomy12040808

Academic Editor: Caterina Morcia

Received: 23 February 2022
Accepted: 26 March 2022
Published: 27 March 2022

Publisher's Note: MDPI stays neutral with regard to jurisdictional claims in published maps and institutional affiliations.

1. Introduction

Rice (*Oryza sativa* L.), being an important cereal crop in the Asian-Pacific region, accounts for a staple food resource for around four billion people worldwide [1]. Nevertheless, like other plants, the growth of rice is continuously threatened by pathogens and herbivore insects during its entire growth cycle and thus leads to an estimated annual loss of 10–30% of the total rice yield [2,3]. Among 20 kinds of known serious paddy pests, brown planthopper (BPH, *Nilaparvata lugens* Stål), is a migratory and monophagous one and becomes the most destructive pest of rice in recent years [4,5]. BPH has sharp, elongated mouthparts that penetrate plant cells and sucks the phloem sap from rice leaf sheath, during which process viral diseases are also transmitted [6]. Serious BPH feeding causes the complete dying of rice, leading to a phenomenon called "hopperburn" [7]. In China, the BPH-infested rice area was estimated at over 25 million hectares, resulting in a rice yield loss of 2.7 million tons between 2005 and 2008 [8,9]. In Asia, the economic loss caused by BPH could reach more than 300 million dollars per year [10].

In agricultural practice, BPH management is still depending heavily on chemical pesticides, which not only cause severe environment pollution and food safety concerns, but also induce increased BPH resistance to chemical pesticides [11]. Therefore, breeding and deploying of resistant cultivars have been the most economically effective and environmental-friendly approach to BPH management [12]. Hence, the identification of BPH resistance genes and the dissection of the underlying resistance mechanism are critical for their successful use in rice breeding. To date, around 40 BPH resistance genes have been identified in the traditional cultivated rice varieties and wild rice species, most of which are usually present in clusters on chromosomes 3, 4, 6, and 12 [13]. Among them, *Bph14*, *Bph3/Bph15*, *Bph26/bph2*, *bph29*, *bph7/Bph9/Bph10/Bph21*, *Bph18/Bph1*, *Bph32*, *Bph6*, *Bph30/Bph40*, and *Bph37* have been isolated and characterized via a map-based cloning approach [14–25], which provide useful targets for marker-assisted selection (MAS) breeding.

However, the resistance of varieties carrying single BPH resistance gene is easy to be quickly broken down due to the rapid adaptation of BPH or evolution of new biotypes [26]. For example, IR26, the first resistant variety with single resistant gene *Bph1*, was released by the International Rice Research Institute (IRRI) in 1973. However, by 1975, IR26 and other *Bph1*-containing resistant varieties were incapable of reducing BPH damage due to the development of BPH biotype II [27]. In 1976, IRRI released IR36 and other resistant varieties harboring *bph2*. However, a few years later, these varieties were also adapted by the new BPH population [28,29]. Therefore, it is still urgent to explore more new broad-spectrum BPH resistance genes and integrate them into rice cultivars to combat the new virulent BPH populations and ultimately achieve durable and broad-spectrum resistance.

IRGC 8678, a Bangladesh indica rice cultivar, was reported to be resistant to BPH biotype 3 [30], but the underlying BPH resistance gene(s) have not been identified yet. In this study, we revealed that IRGC 8678 was highly resistant to the current BPH population of China, the more destructive Bangladesh type. We identified and mapped a new BPH-resistant gene, namely *Bph43*, in IRGC 8678. A highly resistant near-isogenic line (NIL-*Bph43*-9311) was developed by the introgression of *Bph43* into the elite restorer line 9311 through MAS. We also performed comparative transcriptome analysis and explored early defense responsive genes and pathways underlying *Bph43*. Our work identified a new BPH-resistant gene and demonstrated that *Bph43* can be deployed as a valuable donor in BPH resistance breeding programs.

2. Materials and Methods

2.1. Plant Materials, Mapping Populations, and NIL-Bph43-9311 Construction

IRGC 8678, a Bangladesh indica rice cultivar, was previously found to be resistant to BPH biotype III [30]. We obtained IRGC 8678 from the IRRI and crossed it with a BPH-susceptible Chinese elite indica rice cultivar 9311. The resulted F_1 was self-pollinated to generate F_2 mapping population and the corresponding F_3 families for genetic analysis and gene mapping. At the same time, 9311/IRGC 8678 F_1 plant was backcrossed with 9311 to generate the BC_1F_1 population. The heterozygous BC_1F_1 plants showing the resistance to BPH were selected to generate BC_1F_2 populations for further mapping. The near isogenic lines of *Bph43* in 9311 background was developed by the successive backcrossing of the 9311/IRGC 8678 F_1 with 9311. During this process, the gene-linked markers 16–22 and 16–30 flanking the *Bph43* locus were used to select plants with heterozygous *Bph43* from each backcrossed populations for the next step of backcrossing. Finally, one BC_3F_2 individual carrying homozygous *Bph43* and with the least amount of genetic background of IRGC 8678 was selected and designated as NIL-*Bph43*-9311. The detailed procedure for the development of mapping populations and NIL-*Bph43*-9311 is illustrated in Supplementary Figure S1.

2.2. BPH Insects and Evaluation of BPH Resistance

The BPH insects used in this study were collected from rice fields and maintained on the susceptible cultivar Taichung Native 1 (TN1) in a greenhouse at Wuhan University as previously described [20].

For gene mapping, a seedling bulk test for BPH resistance evaluation was performed on the F_3 families as previously described [30,31]. Briefly, approximately 20 seeds per line were sown in a plastic box in 15 cm long rows, with 2.5 cm between rows. A total of two rows of IRGC 8678, three rows of 9311, and three rows of TN1 were randomly sown among the F_3 families as controls. The seedlings were thinned to 15 plants per row on 7 days after sowing. At the third-leaf stage, the seedlings were infested with eight 2nd to 3rd instar BPH nymphs per seedling. The plastic box was completely enclosed in a fine mesh cage that allowed the light to shine through but prevented the escape of the BPH insects. When all of the seedlings of susceptible control TN1 died, the plants were examined and each seedling was given a resistance score of 0 (none damage, none of the leaves shrank and the plant was healthy), 1 (very slight damage or one leaf was yellowing), 3 (one to two leaves were yellowing, or one leaf shrank), 5 (one to two leaves shrank, or one leaf shriveled), 7 (three to four leaves shrank or two to four leaves shriveled, and the plant was still alive), or 9 (the plant died) based on the modified Standard Evaluation System for Rice [30,31]. A lower resistance score represents a higher BPH resistance level. The resistance score of each F_2 individual was then quantified with Microsoft Excel (ME) as the weighted average of the scores of its corresponding F_3 families. The experiments were conducted with at least three biologically independent replicates.

To verify the location of *Bph43*, a total of 320 BC_2F_2 seeds derived from heterozygous *Bph43* BC_2F_1 plant were sown in a plastic box, surrounded with susceptible 9311 and subjected to the BPH resistance assay. When all of the seedlings of 9311 died, each BC_2F_2 individual was examined and given a resistance score.

The performance of BPH insects on rice plants was evaluated as previously described [20]. In the two-host choice test, one 9311 and one NIL-*Bph43*-9311 plants were grown in the same plastic cup (10 cm in diameter, 15 cm in height). At the four-leaf stage, the cup was covered with a light-transmitting nylon mesh, and 20 2nd to 3rd-instar BPH nymphs were released in the cup. The numbers of BPH insects that settled on each plant were recorded at 3, 6, 24, 48, 72, 96, 120, 144, and 168 h after release. The experiments were repeated at least 20 times. The two-host choice test is an indicator of the BPH antixenotic factor.

For BPH weight gain and honeydew excretion assay, newly emerged female adults were weighed, individually enclosed in a pre-weighed Parafilm sachet and attached to the leaf sheath of the rice plant. After 2 days of insect infestation, both the sachets and BPH insects were weighed again. The weight difference of BPH insects was recorded as the weight gain of BPH, and the weight difference of the Parafilm sachet was recorded as honeydew excretion. At least 15 BPH insects were used for analysis.

2.3. DNA Extraction and Gene Mapping

Total genomic DNA was extracted from fresh rice leaves using modified CTAB protocol [32]. The extracted DNA was dissolved in a $1\times$ TE buffer. According to phenotype of F_2 mapping population, equal amounts of DNA from 11 extremely resistant F_2 plants were mixed to form a resistant bulk DNA pool. Similarly, equal amounts of DNA from 18 extremely susceptible F_2 plants were mixed to form a susceptible bulk DNA pool. The green super rice chip-based bulked segregant analysis method (BSA-chip) was used to map the BPH resistance gene. The DNA pools of the resistant bulk and the susceptible bulk as well as the DNAs of the two parents (9311 and IRGC 8678) were subjected to the green super rice chip GSR 40K array at the Greenfafa (Wuhan, China), according to the Infinium HD Assay Ultra Protocol (available online: http://www.illumina.com/ (accessed on 17 September 2021)).

For the insertion-deletion (InDel) calling and development, the DNAs of the two parents (9311 and IRGC 8678) were deeply ($\sim30\times$) sequenced on the platform of Hiseq X Ten. Low-quality reads were deleted or trimmed by a base-quality Q score with a Phred scale of <20 and a read size of <60 bp from raw data using Perl scripts. The cleaned data were mapped to the reference genome of Nipponbare retrieved from Rice Genome Annotation Project website (available online: http://rice.plantbiology.msu.edu/pub/data/Eukaryotic_Projects/o_sativa/annotation_dbs/pseudomolecules/version_7.0/ (accessed on 20 October 2019)) using Burrows–Wheeler Aligner (BWA) software (version 0.5.8c) [33].

InDels of 20 to 40 bp in length between IRGC 8678 and 9311 were extracted using SAMtools (version 0.1.12a) [34]. The upstream and downstream sequences of the candidate InDel loci on the reference genome Nipponbare were obtained and subjected to PCR primer design using Primer 3. The size of PCR products ranged from 160 to 250 bp, so that the polymorphism between IRGC 8678 and 9311 can be easily identified on a 1.5–2% agarose gel. Primers designed in this study are listed in Supplementary Table S1. PCR was performed as previously described [35], with minor modification for different InDel markers. PCR products were separated on a 1.5–2% agarose gel and stained with ethidium bromide.

2.4. RNA-Seq and Transcriptome Analysis

The seedlings of NIL-*Bph43*-9311 and 9311 plants were infested with BPH nymphs (8 s to third-instar nymphs per plant). At 3 h of BPH infestation, leaf sheaths of the rice plants infected by BPH were collected. The samples were referred to as NIL-3 for NIL-*Bph43*-9311 and 9311-3 for 9311, the number representing 3 h of BPH infestation. At the same time, leaf sheaths of NIL-*Bph43*-9311 and 9311 without BPH infestation were collected as an undamaged control and referred to as NIL-0 and 9311-0, respectively. Two to four biological replicates per treatment with 15 seedlings per replicate were used for RNA-seq analysis. Total RNA was isolated using an RNAprep Pure Plant Kit (Tiangen Biotech (Beijing) Co., Ltd., Beijing, China) following the manufacturer's instructions. Concentrations of RNA were checked and integrity was verified on an Agilent 2100 Bioanalyzer (Agilent Technologies, Santa Clara, CA, USA). The cDNA library for each sample was constructed using a TruSeq Stranded mRNA LT Sample Prep Kit (Illumina, San Diego, CA, USA) and quantified on a 150 bp paired-end run by Agilent2100 and sequenced by illumina novascq 6000 (Illumina, San Diego, CA, USA). Clean reads were obtained after the removal of adaptors, low-quality reads and reads with >5% unknown nucleotides with Trimmomatic software [36] and mapped on the rice genome of the cultivar Shuhui498 (available online: http://www.mbkbase.org/R498/ (accessed on 29 October 2021)) using the Hisat2 (v2.2.1.0) [37]. Gene counts were obtained by HTseq, and gene expression was determined using the fragments per kilobase of the exon model per million mapped fragments (RPKM) method [38]. Differentially expressed genes (DEGs) were filtered by DESeq2 after significance. *p*-values were performed at absolute values of log2FC of ≥ 1 ($p < 0.05$) [39]. The gene ontology (GO) enrichment and Kyoto Encyclopedia of Genes and Genomes (KEGG) pathway enrichment analysis of DEGs were performed separately using R based on the hypergeometric distribution [40,41]. The summary of RNA-Seq data is presented in Supplementary Table S2. The Pearson's correlation coefficients for all tested samples are shown in Supplementary Figure S2.

2.5. Quantitative Real-Time PCR (qRT-PCR) Analysis

qRT-PCR was used to validate the DEGs analysis results. All of the rice samples used for qRT-PCR verification were prepared as described above. Total RNAs were isolated from rice plants using TRIzol reagent (TaKaRa) and then converted into first-strand cDNA using PrimeScript™ RT reagent kits accompanied with gDNA Eraser (TaKaRa, code no. RR047A) according to the manufacturer's instructions. Gene expression was analyzed by qRT-PCR using SYBR green supermixes and CFX96 Real-Time System (Bio-Rad). Primers are listed in Supplementary Table S1. The expression levels of genes were calculated with the $2^{-\Delta\Delta C}$ (t) method using CFX Manager Software 2.1. Being stably expressed in rice various organs and developmental stages, *OsAct1* (AB047313) was used as the internal control for qRT-PCR [42]. Each experiment was performed in three biological replicates and three technical replicates.

2.6. Statistical Analysis

The *t*-test was used to examine the difference between two groups. One-way ANOVA was used to compare multiple samples. Statistical tests were conducted using the software GraphPad Prism 7.

3. Results

3.1. Genetic Analysis of BPH Resistance

IRGC 8678, which was previously reported to be resistant to BPH biotype III [30], exhibited high resistance to the current BPH population of China, the more destructive Bangladesh type [43]. The average resistance scores of IRGC 8678 were 1.15 in two independent tests (Figure 1A), suggesting that IRGC 8678 was highly resistant to BPH and might be a useful donor for BPH resistance breeding programs. We further investigated the performance of BPH fed on IRGC 8678 and 9311 plants. Compared to those on 9311 plants, BPH insects fed on IRGC 8678 showed a significantly lower weight gain, which was 0.1808 mg on average (Figure 1B), indicating a strong antibiosis effect on BPH.

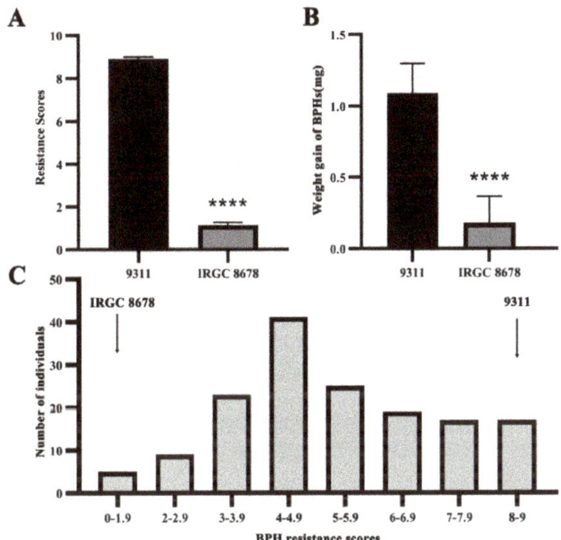

Figure 1. Genetic analysis of brown planthopper (BPH) resistance in IRGC 8678. (**A**) BPH resistance scores of 9311 and IRGC 8678; (**B**) weight gains of BPHs fed on 9311 and IRGC 8678 for 48 h. All data are means ± the standard error of the mean (SEM). **** indicates a significant difference at $p < 0.0001$ by t-test; (**C**) frequency distributions of the BPH resistance scores in the F_2 population derived from 9311 × IRGC 8678.

To explore the genetic basis of BPH resistance in IRGC 8678, we evaluated BPH resistance scores of the F_2 mapping population, which showed a continuous distribution with an apparent valley bottom between 7.0 and 7.9 (Figure 1C). According to a previous BPH resistance-scoring criterion [35], we categorized the plants with resistance scores below 6.99 as resistant plants, whereas those with resistance scores above 7.00 were considered as susceptible ones. The segregation of the resistant to susceptible plants was found to be 122:34, which was in agreement with a 3:1 ratio (122:34; $\chi^2_c = 0.504 < \chi^2_{0.05,1} = 3.84$), demonstrating that BPH resistance of IRGC 8678 was controlled by a major dominant gene.

3.2. Molecular Mapping of Bph43

To map the BPH resistance gene in IRGC 8678, we performed a bulked segregant analysis (BSA) analysis. According to the phenotype of F_2 population, 11 extremely resistant plants and 18 extremely susceptible plants were selected to prepare two contrasting bulks, respectively. The green super rice chip GSR 40K was then used to determine the region containing putative BPH resistance genes. It was revealed that there was only one contiguous region ranging from 8.0 to 18.0 Mb on chromosome 11 significantly differentiating between the resistant and susceptible bulks (Figure 2A), suggesting that the BPH-resistant gene in IRGC 8678 is located in this region. We designed this gene as *Bph43*.

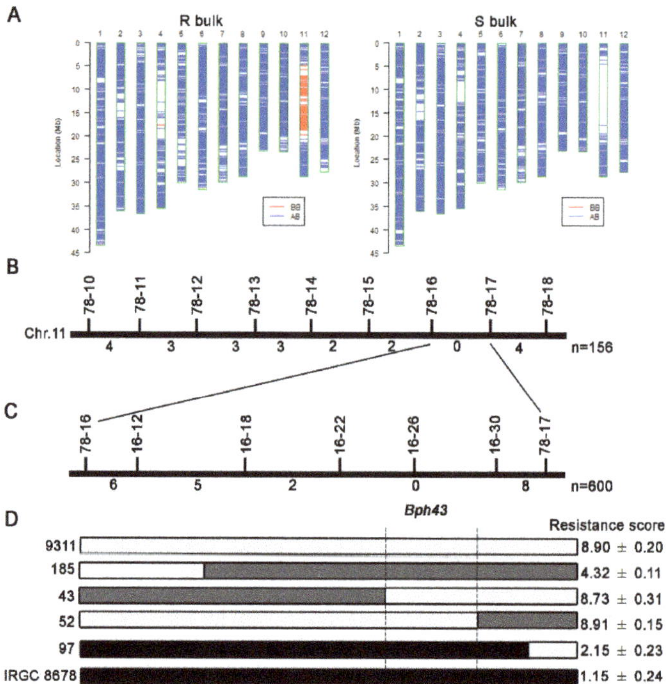

Figure 2. Molecular mapping of *Bph43*. (**A**) Mapping of *Bph43* on rice chromosome 11 using bulked segregant analysis (BSA)-based rice chip GSR 40K analysis. Red lines indicate the genetic background of IRGC 8678, while white ones indicate that of 9311. Blue lines indicate the heterozygous genomic fragments between IRGC 8678 and 9311. R, resistant; S, susceptible; (**B**) mapping of *Bph43* on the genomic region flanked by markers 78–16 and 78–17; (**C**) physical map of marker intervals 78–16 and 78–17. The numbers below the line in (**A,B**) indicate the numbers of recombinants between adjacent markers; (**D**) graphical genotypes and resistance phenotypes of the recombinants. The black, white, and grey bars denote the genotypes of IRGC 8678 homozygotes, 9311 homozygotes, and heterozygotes, respectively.

To detect the exact location of *Bph43* on chromosome 11, we developed 13 polymorphic InDel markers on chromosome 11 and used them to determine the genotypes of 156 F_2 plants with evaluated BPH resistance scores, including those 11 extremely resistant plants and 18 extremely susceptible plants used for BSA-chip analysis. Consequently, *Bph43* was primarily mapped to the 16.2–17.6 Mb region on chromosome 11 flanked by markers 78–16 and 78–17 (Figure 2B). A further analysis of 600 BC_1F_2 plants using the flanking markers 78–16 and 78–17 identified 21 recombinants. These recombinants were genotyped with polymorphic InDel markers within the mapping interval (Figure 2C). According to the genotypes and BPH resistance phenotypes of the recombinants, *Bph43* was located between InDel markers 16–22 and 16–30 (Figure 2D), within intervals of 350 and 380 kb according to R498 and Nipponbare reference genomes, respectively. Putative genes were predicted in the corresponding candidate regions of R498 and Nipponbare reference genomes by using a rice genome annotation project database [44]. A gene cluster encoding putative nucleotide-binding domain leucine-rich repeat-containing (NBS-LRR) proteins and LRR family proteins was identified in *Bph43* mapping regions of both R498 and Nipponbare reference genomes (Supplementary Table S3). Interesting, one of them was found to be specifically induced in NIL-*Bph43*-9311 upon BPH infestation as revealed by RNA-Seq (Supplementary Table S3). As most BPH resistance genes isolated to date encode NBS-LRR proteins, we considered these putative genes encoding NBS-LRR and LRR family proteins as the candidates for *Bph43*.

3.3. Verification of Bph43 in BC$_2$F$_2$ Populations

To confirm the location of *Bph43*, BC$_2$F$_1$ plants that were heterozygous between InDel markers 16–22 and 16–30 were selected to produce BC$_2$F$_2$ generations. BPH resistance evaluation revealed 236 and 84 seedlings were resistant and susceptible to BPH, respectively, and the segregation of the resistant to susceptible plants was in agreement with a 3:1 ratio (236:84, $\chi^2_c = 0.267 < \chi^2_{0.05,1} = 3.84$). We further genotyped these surviving resistant 236 BC$_2$F$_2$ plants with InDel markers 16–22 and 16–30. All of them contained either homozygous or heterozygous alleles from the resistant parent IRGC 8678, verifying the existence of *Bph43* on the region between markers 16–22 and 16–30.

3.4. Development and Characterization of NIL-Bph43-9311

9311, an elite restorer parent for hybrids in China, is highly susceptible to BPH. To improve its BPH resistance, 9311 was used as the recurrent parent to backcross with IRGC 8678 (Supplementary Figure S1). In the process, the flanking markers 16–22 and 16–30 tightly linked to *Bph43* locus were used to select the positive progenies for continuous backcrossing. Simultaneously, background selections using rice chip GSR 40K were conducted. One BC$_3$F$_1$ plant containing a heterozygous *Bph43* region and the least amount of genetic background of IRGC 8678 was selected to produce BC$_3$F$_2$ populations. Finally, one line homozygous for *Bph43* locus and morphologically similar to 9311 was selected as NIL-*Bph43*-9311 (Supplementary Figure S3). Thus, *Bph43* has been successfully introgressed into 9311 through MAS.

We further characterized BPH resistance of NIL-*Bph43*-9311. NIL-*Bph43*-9311 plants that were homozygous or heterozygous for *Bph43* exhibited high resistance to BPH, and no significant difference was detected between *Bph43*-heterozygous and -homozygous plants in terms of resistance scores (Figure 3A,B). The results suggested that *Bph43* is valuable in hybrid rice breeding. BPH insects fed on NIL-*Bph43*-9311 plants showed a significantly lower honeydew excretion and weight gain than those of 9311 (Figure 3C,D), indicating that *Bph43* conferred strong antibiosis effects on BPH. In two-host choice tests, many more BPH insects preferred to settle on 9311 compared with on NIL-*Bph43*-9311 plants (Figure 3E), suggesting that *Bph43* also had a stronger antixenotic effect on BPH.

3.5. Comparative Transcriptome Analysis of NIL-Bph43-9311 and 9311 Underlying BPH Infestation

To further understand the molecular mechanism underlying BPH resistance conferred by *Bph43*, RNA-Seq analysis was performed using the stems of NIL-*Bph43*-9311 and 9311 at the early stage (3 h) of BPH infestation (Supplementary Table S2 and Supplementary Figure S2). One hundred and ninety constitutive DEGs between NIL-*Bph43*-9311 and 9311 without BPH infestation were identified (Supplementary Table S4) and significantly enriched in the defense response, defense response signaling pathway (resistance gene-dependent), plant-type hypersensitive response, and so on (Supplementary Table S5 and Supplementary Figure S4A). At 3 h after BPH infestation, a higher number of DEGs (449 vs. 281 DEGs) were identified in NIL-*Bph43*-9311 plants than in 9311, with 208 down- and 241 upregulated, respectively (Supplementary Table S4). These DEGs were mainly enriched in the regulation of the jasmonic acid mediated signaling pathway, monoterpene biosynthetic process, terpenoid biosynthetic process, regulation of defense response, and so on (Supplementary Table S5 and Supplementary Figure S4C). Upregulated DEGs in NIL-*Bph43*-9311 and 9311 upon BPH infestation were further analyzed. In total, there were 47 upregulated DEGs (G2) shared by NIL-*Bph43*-9311 and 9311, 194 upregulated DEGs (G1) unique to

NIL-*Bph43*-9311, and 183 upregulated DEGs (G3) unique to 9311 (Figure 4A and Supplementary Table S6). We then focused on analyzing DEGs in G1, as these DEGs are putatively involved in the molecular mechanism underlying *Bph43*-mediated resistance. GO enrichment analysis revealed that G1 was significantly enriched in the root development, defense response and defense response signaling pathway (resistance-gene-dependent), hydrogen peroxide catabolic process, plant-type hypersensitive response, and so on (Figure 4B and Supplementary Table S6).

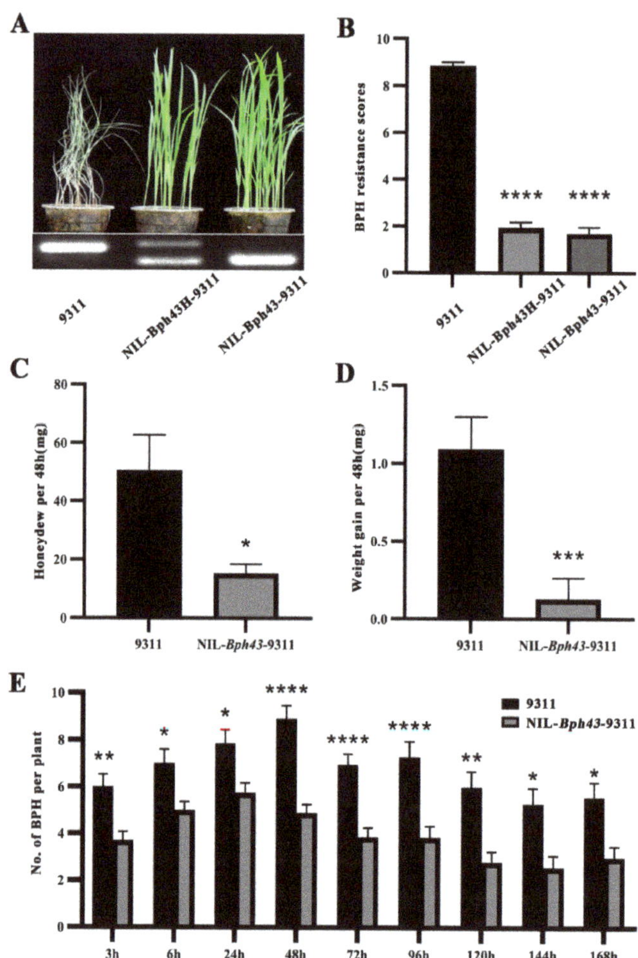

Figure 3. Characterization of BPH resistance of NIL-*Bph43*-9311: (**A**) seedling resistance tests of heterozygous and homozygous NIL-*Bph43*-9311 plants. The PCR band patterns amplified with insertion-deletion (InDel) markers 16–26 showed that NIL-*Bph43*-9311 was homozygous for *Bph43* and heterozygous for NIL-*Bph43H*-9311 and 9311 contained no *Bph43*; (**B**) BPH-resistance scores of 9311, NIL-*Bph43H*-9311, and NIL-*Bph43*-9311; (**C**) honeydew excretion; and (**D**) weight gains of BPH feeding on NIL-*Bph43*-9311 and 9311 for 2 days; (**E**) numbers of BPH insects settled on NIL-*Bph43*-9311 and 9311 in the two-host choice test. Data represent means ± SEM of 20 replicates. The asterisks indicate significant differences (*, $p < 0.05$; **, $p < 0.01$; ***, $p < 0.001$; ****, $p < 0.0001$; Student's *t*-test).

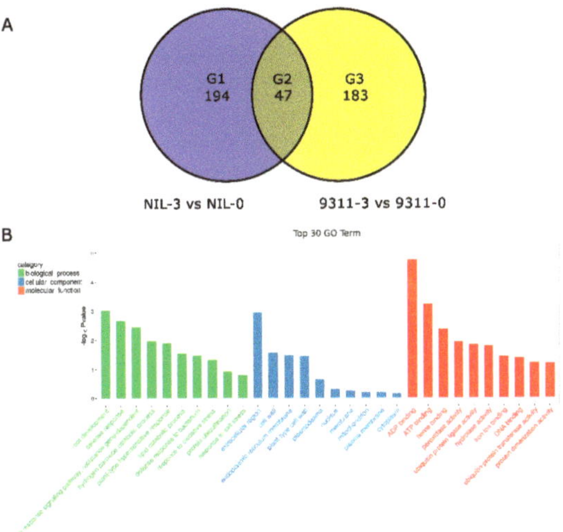

Figure 4. Gene ontology (GO) annotation of upregulated differentially expressed genes (DEGs): (**A**) Venn analysis of upregulated DEGs between NIL-*Bph43*-9311 and 9311; (**B**) top 30 GO terms of specific upregulated DEGs in NIL-*Bph43*-9311. 9311-0 and 9311-3 refer to 9311 sampled at 0 and 3 h of BPH infestation, respectively, and NIL-0 and NIL-3 refer to NIL-*Bph43*-9311 plants sampled at 0 and 3 h of BPH infestation, respectively.

To validate the RNA-Seq results, three DEGs of G2 upregulated in both NIL-*Bph43*-9311 and 9311 and five DEGs of G1 upregulated unique to NIL-*Bph43*-9311 were randomly chosen for the validation of the transcriptomic data using qRT-PCR. The expression profiles of eight genes in qRT-PCR were consistent with the RNA-Seq (Figure 5), confirming the accuracy and reproducibility of the RNA-Seq in the present study.

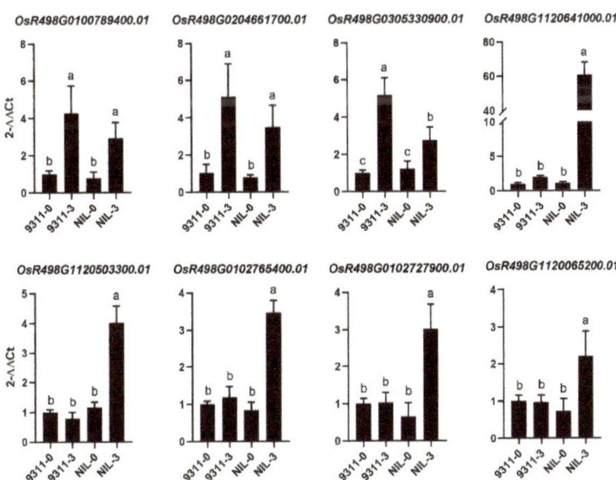

Figure 5. Quantitative real-time PCR (qRT-PCR) validation of the selected DEGs in 9311 and NIL-*Bph43*-9311. OsAct1 was used as an internal control. The relative expression levels of each gene were measured using the $2^{-\Delta\Delta CT}$ method. Data represent the means (three biologically independent experiments for gene expression) $\pm$ SEM. Different letters above the bars indicate significant differences ($p < 0.05$).

4. Discussion

BPH has become the most destructive pest of rice, resulting in a huge annual rice yield loss [4,5]. Breeding and deploying of resistant cultivars is the most effective, economical, and environment-friendly solution for BPH control [12]. So far, around 40 BPH resistance genes have been identified, and 17 of them have been isolated via a map-based cloning approach [13–25]. Most BPH resistance genes are clustered on specific chromosomes. For example, *Bph1*, *bph2*, *bph7*, *Bph9*, *Bph10*, *Bph18*, *Bph21*, and *Bph26* are clustered on chromosome 12L [20,45–49]. Twelve genes are clustered in three regions on chromosome 4 (*Bph30* and *Bph33* in a 0.91–0.97 Mb region; *Bph3/17*, *Bph12*, *Bph15*, *Bph20(t)*, and *bph22(t)* in a 4.1–8.9 Mb region; and *Bph6*, *bph18(t)*, *Bph27*, *Bph27(t)*, and *Bph34* in a 19.1–25.0 Mb), whereas *Bph3*, *bph4*, *bph25*, *bph29*, and *Bph32* are located in a 0.2–1.7 Mb region on chromosome 6 [16,19,50–52]. In the present study, *Bph43* was mapped to a region of ~380 kb on chromosome 11 flanked by InDel markers 16–22 and 16–30. Previously, *Bph28(t)* in rice variety DV85 was physically defined to an interval of 64.8 kb on chromosome 11 [53]. A protein transport Sec24-like gene (LOC_Os11g29200), a tetratricopeptide-like helical domain-containing protein gene (LOC_Os11g29230), a PHD zinc finger (LOC_Os11g29240), and three hypothetical protein genes are predicted in the interval [53]. *Bph28(t)* was located in the region of 16.92–16.99 Mb on chromosome 11 according to Nipponbare genome sequence. The position of *Bph43* was outside this interval. Thus, *Bph43* should be different from *Bph28(t)*. It is reasonable to conclude that *Bph43* is a novel major gene for BPH resistance.

It has been proposed that a large number of BPH resistant varieties come from South Asia [27,30,54]. In this study, IRGC 8678, which is originated from Bangladesh and reported to be resistant to BPH biotype III [30], also showed high resistance to the current BPH population of China, the more destructive Bangladesh type (Figure 1A,B). These results implied that *Bph43* in IRGC 8678 could be a potential durable and broad-spectrum resistance gene. Generally, rice may employ three resistance mechanisms against BPH, including antixenosis, antibiosis, and tolerance [55]. We have demonstrated that *Bph43* in IRGC 8678 confers strong antibiosis and antixenosis effects on BPH. Meanwhile, *Bph43* is a dominant gene, as plants homozygous or heterozygous for *Bph43* both exhibited high resistance to BPH at a similar level (Figure 3A,B). These characteristics make *Bph43* a valuable candidate for hybrid rice breeding and production.

Most BPH resistance genes isolated to date encode nucleotide-binding site and leucine-rich repeat receptors (NLRs) proteins [13–25]. For example, *Bph14* and *Bph9* and its alleles encode classical NLR proteins [17,20]. *Bph6* and *Bph30* encode novel atypical LRR proteins [23,24]. We identified a gene cluster encoding putative NLR and LRR proteins in the *Bph43* mapping region. One of them was specifically induced in NIL-*Bph43*-9311 (Supplementary Table S3). We considered these NBS-LRR and LRR family proteins encoding genes as the candidates for *Bph43*. It has been found that *Bph3* in Rathu Heenati (RH) comprises a cluster of three genes encoding plasma membrane–localized lectin receptor kinases (OsLecRK1–OsLecRK3) which function together to confer broad-spectrum and durable insect resistance in rice [21]. *Bph43* may function in a similar way. Whether these genes work alone or function together requires further investigation.

Transcriptome analysis is an effective way to explore rice–BPH interactions [43,56]. In this study, we also used RNA-Seq to explore the molecular mechanism underlying *Bph43*-mediated resistance at the early stage of BPH feeding. Genes related to defense pathways were enriched in un-infested NIL-*Bph43*-9311 plants, suggesting that the plants might be in a primed state. After BPH infestation, NIL-*Bph43*-9311 initiated more intensive and prompt defense responses. Comparative analysis revealed that there were 194 upregulated DEGs unique to NIL-*Bph43*-9311 upon BPH infestation, significantly enriched in the root development, defense response and defense response signaling pathway (resistance gene-dependent), hydrogen peroxide catabolic process, and plant-type hypersensitive response. These genes are putatively involved in BPH resistance mechanism mediated by *Bph43*.

5. Conclusions

Due to the rapid adaptation of BPH or evolution of new biotypes, it is still urgent to explore and utilize more new broad-spectrum BPH resistance genes. In the current study, we explored the genetic basis of BPH resistance of rice cultivar IRGC 8678. We identified and mapped a new resistance gene *Bph43* to a region of ~380 kb on chromosome 11. *Bph43* was successfully introgressed into elite restorer line 9311 through MAS. The resulted near isogenic lines NIL-*Bph43*-9311 had strong antibiosis and antixenosis effects on BPH. Comparative transcriptome analysis revealed genes related to the defense response and resistance gene-dependent signaling pathway were significantly and uniquely enriched in *Bph43* plants and might be the underlying molecular mechanism for BPH resistance. Our work provides a valuable BPH resistance gene for rice breeding programs.

Supplementary Materials: The following supporting information can be downloaded at: https://www.mdpi.com/article/10.3390/agronomy12040808/s1, Table S1: Primers used in this study. Table S2: Summary of RNA-Seq data. Table S3: Candidate genes predicted in the reference genomes and their expression analysis by RNA-Seq. Table S4: Differentially expressed genes (DEGs) in response to BPH feeding between NIL-*Bph43*-9311 and 9311. Table S5: Gene ontology (GO) annotations of DEGs in response to BPH feeding between NIL-*Bph43*-9311 and 9311. Table S6: Go annotations of upregulated DEGs in response to BPH feeding between NIL-*Bph43*-9311 and 9311. Figure S1: Strategy used to develop NIL-*Bph43*-9311. Figure S2: FPKM box diagram of all tested samples and heatmap of Pearson's correlation coefficients for all tested samples. Figure S3: Genetic background assay of *NIL-Bph43-9311* with rice chip GSR 40K array. Figure S4: Enriched GO annotations of DEGs in response to BPH feeding between NIL-*Bph43*-9311 and 9311.

Author Contributions: R.C. and G.H. conceived and designed the experiments. J.K., X.A., K.Y., S.M., Y.Q., Y.H., B.D. and L.Z. performed the experiments. J.K., X.A. and R.C. analyzed the data. J.K. and R.C. wrote and revised the manuscript. All authors have read and agreed to the published version of the manuscript.

Funding: This research was supported by the National Natural Science Foundation of China (NSFC) (31871598), Shenzhen Fundamental Research Program (JCYJ20180302173435080), and Hubei Hongshan Laboratory (2021hszd005).

Data Availability Statement: Transcriptome datasets are available in the National Center for Biotechnology Information under PRJNA805378.

Acknowledgments: We are very grateful to the International Rice Research Institute (IRRI) for providing resistant paddy materials.

Conflicts of Interest: The authors declare no conflict of interest.

References

1. Sasaki, T.; Burr, B. International Rice Genome Sequencing Project: The effort to completely sequence the rice genome. *Curr. Opin. Plant Biol.* **2000**, *3*, 138–142. [CrossRef]
2. Douglas, A.E. Strategies for Enhanced Crop Resistance to Insect Pests. *Annu. Rev. Plant Biol.* **2018**, *69*, 637–660. [CrossRef] [PubMed]
3. Savary, S.; Willocquet, L.; Pethybridge, S.J.; Esker, P.; McRoberts, N.; Nelson, A. The global burden of pathogens and pests on major food crops. *Nat. Ecol. Evol.* **2019**, *3*, 430–439. [CrossRef] [PubMed]
4. Sogawa, K.; Liu, G.J.; Shen, J.H. A review on the hyper-susceptibility of Chinese hybrid rice to insect pests. *Chin. J. Rice Sci.* **2003**, *17*, 23–30. [CrossRef]
5. Brar, D.S.; Virk, P.S.; Jena, K.; Khush, G.S. Breeding for resistance to planthoppers in rice. In *Planthoppers: New Threats to the Sustainability of Intensive Rice Production Systems in Asia*; Heong, K.L., Hardy, B., Eds.; International Rice Research Institute: Los Baños, Philippines, 2009; pp. 401–427.
6. Fujita, D.; Kohli, A.; Horgan, F.G. Rice Resistance to Planthoppers and Leafhoppers. *Crit. Rev. Plant Sci.* **2013**, *32*, 162–191. [CrossRef]
7. Watanabe, T.; Kitagawa, H. Photosynthesis and Translocation of Assimilates in Rice Plants Following Phloem Feeding by the Planthopper Nilaparvata lugens (Homoptera: Delphacidae). *J. Econ. Entomol.* **2000**, *93*, 1192–1198. [CrossRef]
8. Qiu, Y.; Guo, J.; Jing, S.; Zhu, L.; He, G. Development and characterization of japonica rice lines carrying the brown planthopper-resistance genes BPH12 and BPH6. *Theor. Appl. Genet.* **2012**, *124*, 485–494. [CrossRef]

9. Hu, J.; Xiao, C.; He, Y. Recent progress on the genetics and molecular breeding of brown planthopper resistance in rice. *Rice* **2016**, *9*, 30. [CrossRef]

10. Min, S.; Lee, S.W.; Choi, B.R.; Lee, S.H.; Kwon, D.H. Insecticide resistance monitoring and correlation analysis to select appropriate insecticides against Nilaparvata lugens (Stål), a migratory pest in Korea. *J. Asia-Pac. Entomol.* **2014**, *17*, 711–716. [CrossRef]

11. Tanaka, K.; Endo, S.; Kazano, H. Toxicity of insecticides to predators of rice planthoppers:Spiders, the mirid bug and the dryinid wasp. *Appl. Entomol. Zool.* **2000**, *35*, 177–187. [CrossRef]

12. Matsumura, M.; Takeuchi, H.; Satoh, M.; Sanada-Morimura, S.; Thanh, D.V. Current status of insecticide resistance in rice planthoppers in Asia. In *Planthoppers: New Threats to the Sustainability of Intensive Rice Production Systems in Asia*; Heong, K.L., Hardy, B., Eds.; International Rice Research Institute: Los Baños, Philippines, 2009; pp. 233–244.

13. Du, B.; Chen, R.; Guo, J.; He, G. Current understanding of the genomic, genetic, and molecular control of insect resistance in rice. *Mol. Breed.* **2020**, *40*, 24. [CrossRef]

14. Cheng, X.; Wu, Y.; Guo, J.; Du, B.; Chen, R.; Zhu, L.; He, G. A rice lectin receptor-like kinase that is involved in innate immune responses also contributes to seed germination. *Plant J.* **2013**, *76*, 687–698. [CrossRef]

15. Ji, H.; Kim, S.-R.; Kim, Y.-H.; Suh, J.-P.; Park, H.-M.; Sreenivasulu, N.; Misra, G.; Kim, S.-M.; Hechanova, S.L.; Kim, H.; et al. Map-based Cloning and Characterization of the BPH18 Gene from Wild Rice Conferring Resistance to Brown Planthopper (BPH) Insect Pest. *Sci. Rep.* **2016**, *6*, 34376. [CrossRef] [PubMed]

16. Ren, J.; Gao, F.; Wu, X.; Lu, X.; Zeng, L.; Lv, J.; Su, X.; Luo, H.; Ren, G. Bph32, a novel gene encoding an unknown SCR domain-containing protein, confers resistance against the brown planthopper in rice. *Sci. Rep.* **2016**, *6*, 37645. [CrossRef] [PubMed]

17. Du, B.; Zhang, W.; Liu, B.; Hu, J.; Wei, Z.; Shi, Z.; He, R.; Zhu, L.; Chen, R.; Han, B.; et al. Identification and characterization of Bph14, a gene conferring resistance to brown planthopper in rice. *Proc. Natl. Acad. Sci. USA* **2009**, *106*, 22163–22168. [CrossRef]

18. Tamura, Y.; Hattori, M.; Yoshioka, H.; Yoshioka, M.; Takahashi, A.; Wu, J.; Sentoku, N.; Yasui, H. Map-based Cloning and Characterization of a Brown Planthopper Resistance Gene BPH26 from Oryza sativa L. ssp. indica Cultivar ADR52. *Sci. Rep.* **2014**, *4*, 5872. [CrossRef] [PubMed]

19. Wang, Y.; Cao, L.; Zhang, Y.; Cao, C.; Liu, F.; Huang, F.; Qiu, Y.; Li, R.; Luo, X. Map-based cloning and characterization of BPH29, a B3 domain-containing recessive gene conferring brown planthopper resistance in rice. *J. Exp. Bot.* **2015**, *66*, 6035–6045. [CrossRef]

20. Zhao, Y.; Huang, J.; Wang, Z.; Jing, S.; Wang, Y.; Ouyang, Y.; Cai, B.; Xin, X.-F.; Liu, X.; Zhang, C.; et al. Allelic diversity in an NLR gene BPH9 enables rice to combat planthopper variation. *Proc. Natl. Acad. Sci. USA* **2016**, *113*, 12850–12855. [CrossRef]

21. Liu, Y.; Wu, H.; Chen, H.; Liu, Y.; He, J.; Kang, H.; Sun, Z.; Pan, G.; Wang, Q.; Hu, J.; et al. A gene cluster encoding lectin receptor kinases confers broad-spectrum and durable insect resistance in rice. *Nat. Biotechnol.* **2015**, *33*, 301–305. [CrossRef]

22. Jing, S.; Zhao, Y.; Du, B.; Chen, R.; Zhu, L.; He, G. Genomics of interaction between the brown planthopper and rice. *Curr. Opin. Insect Sci.* **2017**, *19*, 82–87. [CrossRef]

23. Guo, J.; Xu, C.; Wu, D.; Zhao, Y.; Qiu, Y.; Wang, X.; Ouyang, Y.; Cai, B.; Liu, X.; Jing, S.; et al. Bph6 encodes an exocyst-localized protein and confers broad resistance to planthoppers in rice. *Nat. Genet.* **2018**, *50*, 297–306. [CrossRef] [PubMed]

24. Shi, S.; Wang, H.; Nie, L.; Tan, D.; Zhou, C.; Zhang, Q.; Li, Y.; Du, B.; Guo, J.; Huang, J.; et al. Bph30 confers resistance to brown planthopper by fortifying sclerenchyma in rice leaf sheaths. *Mol. Plant* **2021**, *14*, 1714–1732. [CrossRef] [PubMed]

25. Zhou, C.; Zhang, Q.; Chen, Y.; Huang, J.; Guo, Q.; Li, Y.; Wang, W.; Qiu, Y.; Guan, W.; Zhang, J.; et al. Balancing selection and wild gene pool contribute to resistance in global rice germplasm against planthopper. *J. Integr. Plant Biol.* **2021**, *63*, 1695–1711. [CrossRef] [PubMed]

26. Jena, K.K.; Kim, S.-M. Current Status of Brown Planthopper (BPH) Resistance and Genetics. *Rice* **2010**, *3*, 161–171. [CrossRef]

27. Khush, G.S.; Coffman, W.R. Genetic Evaluation and Utilization (GEU) program. *Theor. Appl. Genet.* **1977**, *51*, 97–110. [CrossRef]

28. Alam, S.N.; Cohen, M.B. Detection and analysis of QTLs for resistance to the brown planthopper, Nilaparvata lugens, in a doubled-haploid rice population. *Theor. Appl. Genet.* **1998**, *97*, 1370–1379. [CrossRef]

29. Ketipearachchi, Y.; Kaneda, C.; Nakamura, C. Adaptation of the brown planthopper (BPH), Nilaparvata lugens (Stal) (Homoptera: Delphacidae), to BPH resistant rice cultivars carrying bph8 or Bph9. *Appl. Entomol. Zool.* **1998**, *33*, 497–505. [CrossRef]

30. Heinrichs, E.A.; Medrano, F.G.; Rapusas, H.R. *Genetic Evaluation for Insect Resistance in Rice*; International Rice Research Institute: Los Baños, Philippines, 1985.

31. Huang, Z.; He, G.; Shu, L.; Li, X.; Zhang, Q. Identification and mapping of two brown planthopper resistance genes in rice. *Theor. Appl. Genet.* **2001**, *102*, 929–934. [CrossRef]

32. Porebski, S.; Bailey, L.G.; Baum, B.R. Modification of a CTAB DNA extraction protocol for plants containing high polysaccharide and polyphenol components. *Plant Mol. Biol. Report.* **1997**, *15*, 8–15. [CrossRef]

33. Li, H.; Durbin, R. Fast and accurate long-read alignment with Burrows-Wheeler transform. *Bioinformatics* **2010**, *26*, 589–595. [CrossRef]

34. Li, H.; Handsaker, B.; Wysoker, A.; Fennell, T.; Ruan, J.; Homer, N.; Marth, G.; Abecasis, G.; Durbin, R.; Genome Project Data Processing Subgroup. The Sequence Alignment/Map format and SAMtools. *Bioinformatics* **2009**, *25*, 2078–2079. [CrossRef] [PubMed]

35. Qiu, Y.; Guo, J.; Jing, S.; Zhu, L.; He, G. High-resolution mapping of the brown planthopper resistance gene Bph6 in rice and characterizing its resistance in the 9311 and Nipponbare near isogenic backgrounds. *Theor. Appl. Genet.* **2010**, *121*, 1601–1611. [CrossRef] [PubMed]

36. Bolger, A.M.; Lohse, M.; Usadel, B. Trimmomatic: A flexible trimmer for Illumina sequence data. *Bioinformatics* **2014**, *30*, 2114–2120. [CrossRef] [PubMed]

37. Kim, D.; Paggi, J.M.; Park, C.; Bennett, C.; Salzberg, S.L. Graph-based genome alignment and genotyping with HISAT2 and HISAT-genotype. *Nat. Biotechnol.* **2019**, *37*, 907–915. [CrossRef]

38. Anders, S.; Pyl, P.T.; Huber, W. HTSeq—a Python framework to work with high-throughput sequencing data. *Bioinformatics* **2014**, *31*, 166–169. [CrossRef]

39. Love, M.I.; Huber, W.; Anders, S. Moderated estimation of fold change and dispersion for RNA-seq data with DESeq2. *Genome Biol.* **2014**, *15*, 550. [CrossRef]

40. Kanehisa, M.; Araki, M.; Goto, S.; Hattori, M.; Hirakawa, M.; Itoh, M.; Katayama, T.; Kawashima, S.; Okuda, S.; Tokimatsu, T.; et al. KEGG for linking genomes to life and the environment. *Nucleic Acids Res.* **2008**, *36*, D480–D484. [CrossRef]

41. The Gene Ontology Consortium. The Gene Ontology Resource: 20 years and still GOing strong. *Nucleic Acids Res.* **2018**, *47*, D330–D338. [CrossRef]

42. Hu, L.; Wu, Y.; Wu, D.; Rao, W.; Guo, J.; Ma, Y.; Wang, Z.; Shangguan, X.; Wang, H.; Xu, C.; et al. The Coiled-Coil and Nucleotide Binding Domains of BROWN PLANTHOPPER RESISTANCE14 Function in Signaling and Resistance against Planthopper in Rice. *Plant Cell* **2017**, *29*, 3157–3185. [CrossRef]

43. Lv, W.; Du, B.; Shangguan, X.; Zhao, Y.; Pan, Y.; Zhu, L.; He, Y.; He, G. BAC and RNA sequencing reveal the brown planthopper resistance gene BPH15 in a recombination cold spot that mediates a unique defense mechanism. *BMC Genom.* **2014**, *15*, 674. [CrossRef]

44. Kawahara, Y.; de la Bastide, M.; Hamilton, J.P.; Kanamori, H.; McCombie, W.R.; Ouyang, S.; Schwartz, D.C.; Tanaka, T.; Wu, J.; Zhou, S.; et al. Improvement of the Oryza sativa Nipponbare reference genome using next generation sequence and optical map data. *Rice* **2013**, *6*, 4. [CrossRef] [PubMed]

45. Ishii, T.; Brar, D.S.; Multani, D.S.; Khush, G.S. Molecular tagging of genes for brown planthopper resistance and earliness introgressed from Oryza australiensis into cultivated rice, O. sativa. *Genome* **1994**, *37*, 217–221. [CrossRef] [PubMed]

46. Jena, K.K.; Jeung, J.U.; Lee, J.H.; Choi, H.C.; Brar, D.S. High-resolution mapping of a new brown planthopper (BPH) resistance gene, Bph18(t), and marker-assisted selection for BPH resistance in rice (*Oryza sativa* L.). *Theor. Appl. Genet.* **2006**, *112*, 288–297. [CrossRef] [PubMed]

47. Sun, L.H.; Wang, C.M.; Su, C.C.; Liu, Y.Q.; Zhai, H.Q.; Wan, J.M. Mapping and Marker-assisted Selection of a Brown Planthopper Resistance Gene bph2 in Rice (*Oryza sativa* L.). *Acta Genet. Sin.* **2006**, *33*, 717–723. [CrossRef]

48. Rahman, M.L.; Jiang, W.; Chu, S.H.; Qiao, Y.; Ham, T.H.; Woo, M.O.; Lee, J.; Khanam, M.S.; Chin, J.H.; Jeung, J.U.; et al. High-resolution mapping of two rice brown planthopper resistance genes, Bph20(t) and Bph21(t), originating from Oryza minuta. *Theor. Appl. Genet.* **2009**, *119*, 1237–1246. [CrossRef]

49. Myint, K.K.M.; Fujita, D.; Matsumura, M.; Sonoda, T.; Yoshimura, A.; Yasui, H. Mapping and pyramiding of two major genes for resistance to the brown planthopper (Nilaparvata lugens [Stål]) in the rice cultivar ADR52. *Theor. Appl. Genet.* **2012**, *124*, 495–504. [CrossRef]

50. Jairin, J.; Phengrat, K.; Teangdeerith, S.; Vanavichit, A.; Toojinda, T. Mapping of a broad-spectrum brown planthopper resistance gene, Bph3, on rice chromosome 6. *Mol. Breed.* **2007**, *19*, 35–44. [CrossRef]

51. Jairin, J.; Sansen, K.; Wongboon, W.; Kothcharerk, J. Detection of a brown planthopper resistance gene bph4 at the same chromosomal position of Bph3 using two different genetic backgrounds of rice. *Breed. Sci.* **2010**, *60*, 71–75. [CrossRef]

52. Yara, A.; Phi, C.N.; Matsumura, M.; Yoshimura, A.; Yasui, H. Development of near-isogenic lines for BPH25(t) and BPH26(t), which confer resistance to the brown planthopper, Nilaparvata lugens (Stål.) in indica rice 'ADR52'. *Breed. Sci.* **2010**, *60*, 639–647. [CrossRef]

53. Wu, H.; Liu, Y.; He, J.; Liu, Y.; Jiang, L.; Liu, L.; Wang, C.; Cheng, X.; Wan, J. Fine mapping of brown planthopper (Nilaparvata lugens Stål) resistance gene Bph28(t) in rice (*Oryza sativa* L.). *Mol. Breed.* **2014**, *33*, 909–918. [CrossRef]

54. Kaneda, C.; Ito, K.; Ikeda, R. Screening of Rice Cultivars for Resistance to the Brown Planthopper, Nilaparvata lugens Stal., by Three Biotypes. *Jpn. J. Breed.* **1981**, *31*, 141–151. [CrossRef]

55. Cohen, M.B.; Alam, S.N.; Medina, E.B.; Bernal, C.C. Brown planthopper, Nilaparvata lugens, resistance in rice cultivar IR64: Mechanism and role in successful N. lugens management in Central Luzon, Philippines. *Entomol. Exp. Et Appl.* **1997**, *85*, 221–229. [CrossRef]

56. Tan, J.; Wu, Y.; Guo, J.; Li, H.; Zhu, L.; Chen, R.; He, G.; Du, B. A combined microRNA and transcriptome analyses illuminates the resistance response of rice against brown planthopper. *BMC Genom.* **2020**, *21*, 144. [CrossRef] [PubMed]

 *agronomy*

Article

Genome-Wide Comparative Analysis of Transposable Elements by Matrix-TE Method Revealed Indica and Japonica Rice Evolution

Zhiguo Wu [1,*], Wei Xi [1], Zixuan Han [1], Yanhua Wu [1], Yongzhuo Guan [1] and Yuxian Zhu [1,2]

[1] College of Life Sciences, Wuhan University, Wuhan 430072, China; 2020202040099@whu.edu.cn (W.X.); hanzixuanchn@163.com (Z.H.); 17865815412@163.com (Y.W.); gyzcon@163.com (Y.G.); zhuyx@whu.edu.cn (Y.Z.)

[2] Institute for Advanced Studies, Wuhan University, Wuhan 430072, China

* Correspondence: wu.zhiguo@whu.edu.cn

Abstract: Transposons (TEs) are known to change the gene expression and function, and subsequently cause plant speciation and evolution. Nevertheless, efficient and new approaches are required to investigate the role of TEs in the plant genome structural variations. Here, we reported the method named matrix-TE to investigate the differentiation of intact and truncated LTR/TEs comprehensively in *Indica* and *Japonica* rice throughout whole genomes with a special eye on centromeric regions. Six LTR/TE super-families were identified in both *Indica* and *Japonica* rice genomes, and the TE ORF references were extracted by phylogenetic analysis. *Indica* rice specific TE peak P-*Gypsy* and *Japonica* rice specific TE peak P-*Copia* were observed, and were further analyzed by Gaussian probability density function (GPDF) fit. The individual TE peak P-*Gypsy* was observed in centromeric regions of the *Indica* genome. By the matrix-TE method, the divergence of *Indica* and *Japonica* genomes, especially their centromeric regions, mainly resulted from the *Ty3/Gypsy* insertion events at 0.77 Mya. Our data indicate that the optimized matrix-TE approach may be used to specifically analyze the TE content, family evolution, and time of the TE insertions.

Keywords: transposon; rice genome; centromere; divergence; Gaussian distribution

Citation: Wu, Z.; Xi, W.; Han, Z.; Wu, Y.; Guan, Y.; Zhu, Y. Genome-Wide Comparative Analysis of Transposable Elements by Matrix-TE Method Revealed Indica and Japonica Rice Evolution. *Agronomy* **2022**, *12*, 1490. https://doi.org/10.3390/agronomy12071490

Academic Editor: Katherine Steele

Received: 22 May 2022
Accepted: 18 June 2022
Published: 22 June 2022

Publisher's Note: MDPI stays neutral with regard to jurisdictional claims in published maps and institutional affiliations.

1. Introduction

TEs show an importance in both monocotyledonous and dicotyledonous plants, since TEs probably lead to the alterations of plant gene expression and function by introducing mutations both in the coding regions and the regulatory regions [1]. Recent great advances in genomics has gradually made it a reality to systematically investigate the roles of TEs during the evolution of plant genomic structures [2–5]. Long terminal repeat transposable elements (LTR/TEs) are retrotransposons and usually constitute a major part of most plant genomes [6].

Indica and *Japonica* rice are the most widely cultivated subspecies and the genome sequences were successfully assembled and polished in the past two decades, with their genome structure variations studied to interpret the origin and evolution for the *Oryza* genus [7–12]. The rice genomes are known to contain high levels of *Ty1/Copia* and *Ty3/Gypsy* super-family LTR/TEs [4,5]. To date, the *Oryza* species have been evolved for ~15 million years. However, the divergence time between *Indica* and *Japonica* rice was estimated at 0.55 million years ago (Mya) [11,12]. Molecular phylogenetic studies suggested that *Indica* and *Japonica* rice originated independently [11]. Recently, gap-free rice genomes have been assembled and provide a new view of the structure and function of full-length centromeres with LTR/TEs being a major component [4,5]. These advances provide a firm basis on which to study the essential roles that LTR/TEs play during the origin and evolution of rice genomes.

LTR/TEs are found everywhere in many plant genomes, especially in wheat and maize, as more than half of their entire genome sequences are constituted by *Ty1/Copia* and *Ty3/Gypsy* super-families [13–17]. A large number of intact and truncated LTR/TEs were identified in both the wheat D and A subgenomes. Sequential insertion followed by silencing events of TEs over the past three million years were thought to play a significant role during evolution of the hexoploid wheat genome [13,14]. Furthermore, comparative studies of TEs in three maize genomes interpreted the existence of high level variations, even among very related subspecies [15–17], and TE-mediated gene silencing was proposed to be involved in epigenetic regulations during different developmental stages [18].

Most plant TEs cumulate a large amount of nucleotide substitution and truncation events over their evolution time courses [19–25]. The high repetitive nature with a high mutation rate makes it a challenge to accurately analyze TEs during genome assembly and related studies. Previously, a software entitled TRACK-POSON was published to detect TE insertion polymorphisms in rice genomes [26]. Here, we reported a matrix-TE approach to quantitatively and systematically study the intact and truncated LTR/TEs in the genomes of *Indica* and *Japonica* rice. All of the LTR/TE super-families were restricted in one super matrix by their ORFs and sequence identities, then classified and phylogenetically clustered. Individual TE insertion peaks of P-*Gypsy* and P-*Copia* were detected in *Indica* and *Japonica* rice, respectively. The TE peaks were further resolved by Gaussian probability density function (GPDF), as the GPDF model was matched with stochastic nucleotide substitution events, and the time point of the individual TE peak was calculated by the nucleotide substitution ratio *Ks* [3]. We suggest that this is the first characterization of the insertion events for all rice LTR/TE super-families in the genomes of two subspecies, especially in centromeric regions.

2. Materials and Methods

2.1. Indica and Japonica Genomes Used for TE Matrix Generation and GPDF Analysis

Two cultivated rice *Indica* (MH63) and *Japonica* (Nip) genome sequence data were applied to the analysis process. The *Indica* (MH63) genome assembly ID was GWH-BCKY00000000 and the *Japonica* (Nip) genome assembly ID was GCF_001433935.1_IRGSP-1.0 [4]. Intact LTR/TEs of the MH63 and Nip genomes were annotated by LTR_finder software with default parameters.

2.2. LTR/TE ORF Matrix Generation

The ORFs were extracted from the intact LTR/TEs by the Getorf Script in the EMBOSS package with default parameters. Then, the ORFs were sorted by length from long to short, and we discarded ORFs shorter than 1500 bp. The identity matrix for the selected longer ORF sequences was calculated by BioEdit software with default parameters [3].

2.3. Phylogenetic Analysis of TE ORFs

The LTR/TE ORFs with sequence identities greater than 95% in the TE matrix were extracted according to their sequence homology. Then, the MEGA software with the construct neighbor-joining tree method was used to generate the molecular phylogenetic trees of the ORF groups. The LTR/TE ORF sequences on each top branch of the phylogenetic trees were determined as the TE ORF reference sequences of the super-families in the following steps.

2.4. Whole Genome and Centromere Scanning with TE ORF Reference Sequences and GPDF Analysis of Individual TE Peaks

The *Indica* (MH63) and *Japonica* (Nip) whole genomes and centromeric regions were searched by the TE ORF reference sequences with BLASTN software using the parameters: $blastn -db Ricegenomedb -query ref-ORFseq.fa -out ref-ORFseq-blast-Ricegenomedb -evalue 0.00001 -word_size 11 -gapopen 5 -gapextend 2 -penalty -2 -reward 1 -culling_limit 0 -outfmt 7.

Sequences obtained from the above blast search contained intact and truncated TE sequences, and identity distribution curves were generated using individual TE super-family sequences. The individual peaks of the curves were fitted by GPDF and the average nucleotide substitution ratio *Ks* of each TE peak were defined as 2.58σ [3]. The TE insertion time was calculated by the formula: T = *Ks*/2*r*, where *r* is the average nucleotide substitution rate (1.3 × 10^{-8} here) [27].

2.5. TE Insertion Events, Whole Genome, and Centromere Evolution Analysis

The distribution of the single nucleotide polymorphism across *Ty1*/*Copia* and *Ty3*/*Gypsy* ORFs were scanned in Nip and MH63 genomes to calculate the SNP densities. The individual TE peaks were compared with *Indica* and *Japonica* rice genomes as well as the centromeric regions. Different TE insertion events among the whole genomes and centromeric regions were applied to correlate with the differentiation of rice subspecies [28,29].

3. Results

3.1. Development of Matrix-TE Approach Pipeline

Considering the large amount of TEs and the stochastic nucleotide substitutions, we developed the approach matrix-TE to successfully evaluate the TE ORFs, and calculated the *Ks* of the TE insertion events in rice genomes (Figure S1). Whole rice genome sequences were sequentially applied to the LTR_finder, Getorf, and BioEdit scripts to obtain various related data. ORF clusters with an identity over 95% were observed at the diagonal of the TE matrix and were extracted based on sequence identities. The whole genome or the centromeric regions were scanned by the reference sequence with the identity distribution curve fitted by the GPDF model. The individual TE insertion event was analyzed by calculating the *Ks* derived from the GPDF. The matrix-TE approach was used to analyze the most abundant TE super-families, and the TE content at both the whole genome level and in the centromeric regions were subsequently quantified.

3.2. TE Matrix and Cluster Generation for Indica and Japonica Rice Whole Genomes

Compared with that of Nip, MH63 had a slightly bigger genome and higher TE content with both having more annotated intact LTR/TEs and ORFs (Table 1). A total of 1520 Nip ORFs and 2010 MH63 ORFs were discovered from the super matrices, and clusters with identities over 95% observed at the matrix diagonals (Figure 1A,B) were extracted and annotated so that both the Nip and MH63 genomes contained six clusters of LTR/TE super families, named *type1*, *type2*, *typeRT*, *typePHA*, *Ty1*/*Copia*, and *Ty3*/*Gypsy*. Both the Nip and MH63 genomes contained similar numbers of *type1*, *type2*, *typeRT*, and *typePHA* TEs (from 51 to 64), whereas significantly more of the *Ty1*/*Copia* TEs were observed in Nip than in MH63 (comparing 48 in Nip to 18 in the later). In contrast, MH63 contained significantly more *Ty3*/*Gypsy* type TEs than that of Nip (71 in MH63 while there were only 14 in the latter) (Figure 1C,D).

Table 1. Genome size, TE content, and intact TE ORF statistics of two rice genomes.

Species	*Japonica* (**Nip**)	*Indica* (**MH63**)
Genome size (Mb)	380	395
Total TE content (%)	44.1	45.9
Intact LTR/TE	4744	5146
ORFs in matrix	1520	2010

3.3. Phylogenetic Trees and ORF Reference Sequences of TE Clusters

The phylogenetic trees of TE ORFs were constructed with the six TE clusters of the Nip and MH63 genomes separately (Figure 2). With a non-homologous sequence as the root (e.g., using *Ty1*/*Copia* as the root of *Ty3*/*Gypsy* tree and vice versa), the *ref-N-type1* and *ref-M-type1* (Figure 2A), the *ref-N-type2* and *ref-M-type2* (Figure 2B), the *ref-N-typeRT* and *ref-M-typeRT* (Figure 2C), the *ref-N-typePHA* and *ref-M-typePHA* (Figure 2D), the *ref-N-Ty1/Copia* and *ref-M-Ty1/Copia* (Figure 2E), and the *ref-N-Ty3/Gypsy* and *ref-M-Ty3/Gypsy* (Figure 2F) on the top branch of each tree were selected as the reference sequences for the TE ORFs of each Nip or MH63 TE cluster. These reference sequences were potentially active and were probably inserted into rice genomes in recent ages [2,3]. These TE ORF reference sequences were applied to whole genome and centromere scanning in the following steps.

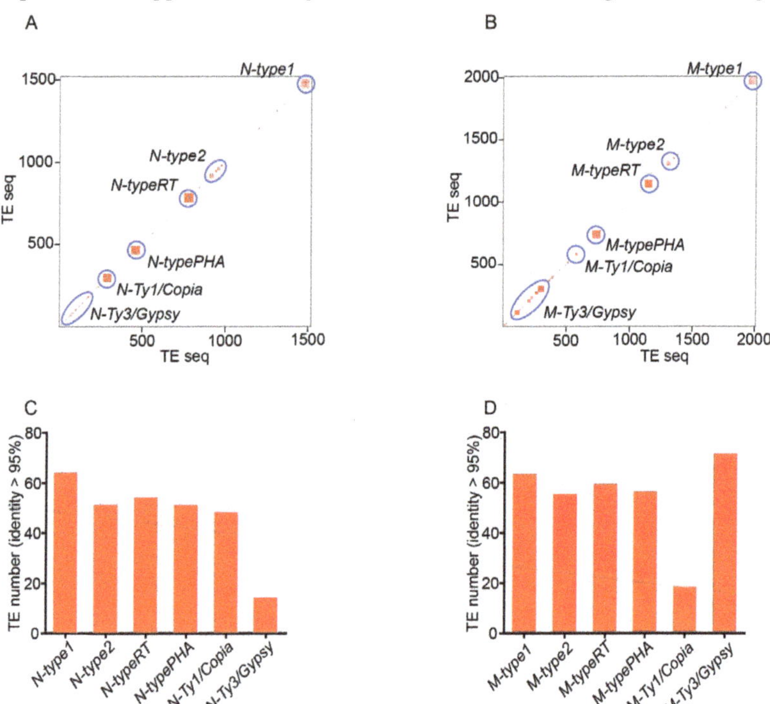

Figure 1. The Nip and MH63 TE ORF matrix and clusters. (**A,B**) Six TE ORF clusters *type1*, *type2*, *typeRT*, *typePHA*, *Ty1/Copia*, and *Ty3/Gypsy* with identities of over 95% observed at the TE matrix diagonals in the Nip and MH63 genomes, respectively, where *N* indicates Nip and *M* indicates MH63. (**C**) TE numbers with identities of over 95% for each cluster in Nip, *N-type1* (64 TEs), *N-type2* (51 TEs), *N-typeRT* (54 TEs), *N-typePHA* (51 TEs), *N-Ty1/Copia* (48 TEs), *N-Ty3/Gypsy* (14 TEs). (**D**) TE numbers with an identity over 95% for each cluster in MH63, *M-type1* (63 TEs), *M-type2* (55 TEs), *M-typeRT* (59 TEs), *M-typePHA* (56 TEs), *M-Ty1/Copia* (18 TEs), and *M-Ty3/Gypsy* (71 TEs).

3.4. Whole Genome and Centromere Scanning by TE ORF Reference Sequences

Both whole genome and the centromeric regions of Nip or MH63 were scanned with the TE ORF reference sequences. Data analysis indicated that the TE ORF contents for the *type1*, *type2*, *typeRT*, and *typePHA* clusters were similar in the Nip or MH63 genomes (Table 2), with similar identity distribution curves and in general quite weak TE hit signals for these four super-families (Figure 3E–L). There were 1479 *Ty1/Copia* ORFs in Nip vs. 316 in MH63, and 6277 *Ty3/Gypsy* ORFs in MH63 compared with 3328 in the Nip genomes. In the Nip centromeres, a total of 57 *Ty1/Copia* and 345 *Ty3/Gypsy* ORFs were observed, while 23 *Ty1/Copia* and 758 *Ty3/Gypsy* ORFs were found in that of MH63 (Table 2), which may indicate the importance of the activities of *Ty1/Copia* and *Ty3/Gypsy* in the evolutions of the Nip and MH63 genomes, respectively. Indeed, sharp peaks of *P-Copia* and *P-Gypsy* were detected in the Nip and MH63 genomes (Figure 3A,B), whereas significantly lower levels of *Ty3/Gypsy* were found in Nip and only some scattered *Ty1/Copia* was revealed in MH63 (Figure 3C,D).

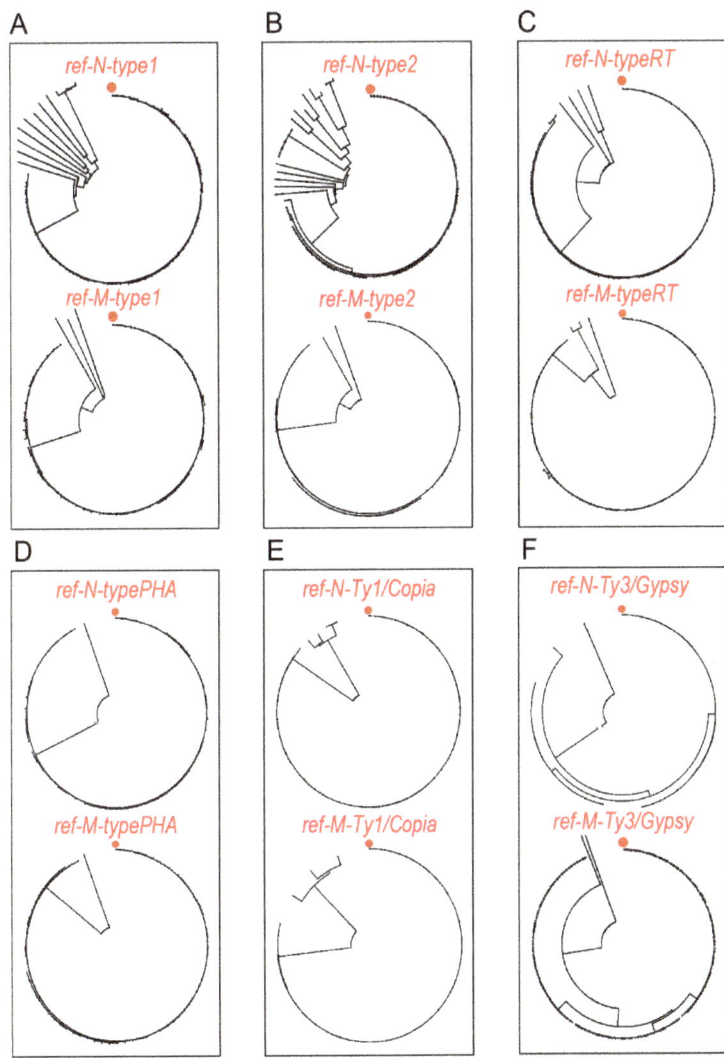

Figure 2. The phylogenetic analysis of the Nip and MH63 TE super-family clusters and ORF reference sequence identifications for each cluster. (**A**) Phylogenetic trees of *type1* TE ORFs for Nip and MH63, the *ref-N-type1* and *ref-M-type1* sequence on the top branch was selected as the reference sequences for the *N-type1* and *M-type1* clusters, respectively. (**B–F**) In the same format as (**A**), for *type2*, *typeRT*, *typePHA*, *Ty1/Copia* and *Ty3/Gypsy*, respectively. The TE ORF reference sequences on the top of the branch for each cluster are shown in red.

Table 2. The whole genome and centromere scanning with six TE ORF reference sequences in MH63 and Nip.

		Genome and Centromere Scanning by TE ORFs					
		Type1	*Type2*	*TypeRT*	*TypePHA*	*Ty1/Copia*	*Ty3/Gypsy*
Number of TE ORFs in genomes	Indica (MH63)	728	1086	998	2379	316	6277
	Japonica (Nip)	820	1039	859	2219	1479	3328
Number of TE ORFs in centromeres	Indica (MH63)	/	/	/	/	23	758
	Japonica (Nip)	/	/	/	/	57	345

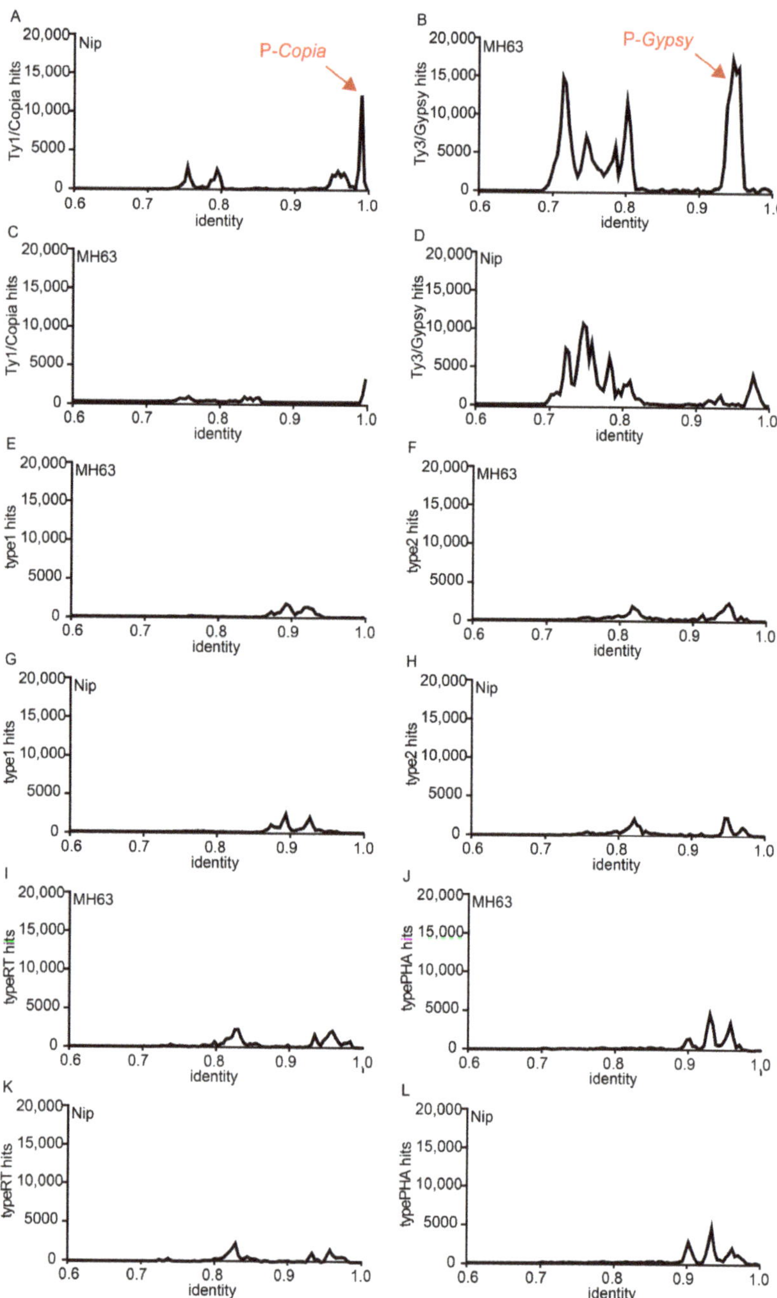

Figure 3. Scanning of the Nip and MH63 whole genomes with the TE ORF reference sequences. (**A**) Nip genome scanning after normalization with *ref-N-Ty1/Copia*. (**B**) MH63 genome scanning after normalization with *ref-M-Ty3/Gypsy*. (**C–L**) in the same format as (**A**), for *Ty1/Copia, Ty3/Gypsy, type1, type2, typeRT* and *typePHA*, respectively. Individual TE peaks P-*Copia* (*p* value = 0.0001) and P-*Gypsy* (*p* value = 0.008) were observed in the Nip and MH63 genomes separately.

3.5. Stochastic SNP Distribution in TE ORFs, and GPDF Analysis of P-Copia and P-Gypsy

Across the *Ty1/Copia* and *Ty3/Gypsy* ORF sequences in both MH63 and Nip, the SNP distribution was observed. A significantly much higher rate of the SNP distributions was found for *Ty3/Gypsy* ORFs than that of *Ty1/Copia* ORFs in both rice genomes (Figure 4A,B). We further produced identity distribution curves for P-*Copia* and P-*Gypsy*, and they fitted well to the mathematical GPDF model (Figure 4C,D). We reported the R square values in the inset of Figure 4C,D, respectively. The nucleotide substitution ratio (*Ks*) of the two peaks were calculated as 2.58σ [3]. The *Ks* value of P-*Copia* was calculated to be 0.0049, while that of P-*Gypsy* was 0.020. The individual P-*Copia* peak representing the *Ty1/Copia* insertion events in the Nip genome were estimated to be 0.19 Mya by GPDF fitting (Figure 4C), and the individual P-*Gypsy* peak representing the *Ty3/Gypsy* insertion events in the MH63 genome were calculated to be at around 0.77 Mya (Figure 4D). The *Ty3/Gypsy* peaks with identity at *Ks* values of 0.7–0.8 were estimated to be inserted in the MH63 genome at 6.3~9.7 Mya (Figure 4D).

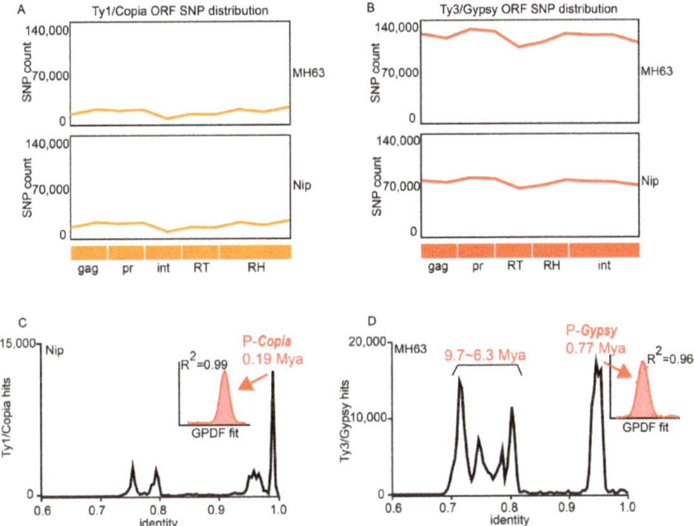

Figure 4. The SNP distributions across *Ty1/Copia* and *Ty3/Gypsy* ORFs, and the GPDF fit of the P-*Copia* and P-*Gypsy* peaks. (**A**) The SNP level distributions of *Ty1/Copia* in the MH63 (upper panel) and Nip (lower panel) genomes. (**B**) The SNP level distributions of *Ty3/Gypsy* in the MH63 (upper panel) and Nip genomes (lower panel). gag, the GAG protein; pr, the protease; int, the integrase; RT, the reverse transcriptase; RH, RNaseH. (**C**) The GPDF fit of the P-*Copia* peak in the Nip genome with an R square value 0.99, and the age of the peak was calculated at around 0.19 Mya. (**D**) The GPDF fit of the P-*Gypsy* peak in MH63 with an R square value of 0.96, and the age of the peak was 0.77 Mya. The peaks with identities of 0.7–0.8 were estimated as *Ty3/Gypsy* insertion events at 9.7~6.3 Mya.

3.6. LTR/TE Analysis in Nip and MH63 Centromeric Regions

Centromere sequences of the MH63 and Nip genomes were extracted from the whole genome data (Tables S1 and S2) [4], and were scanned with the *Ty1/Copia* and *Ty3/Gypsy* ORF reference sequences (Figure 5). No significant *Ty1/Copia* ORF distribution signals were observed in centromeres of MH63 (Figure 5A) or in that of Nip (Figure 5C). However, strong distribution signals for *Ty3/Gypsy* ORFs in the centromeric regions of MH63 (Figure 5B) and Nip (Figure 5D), although the 0.77 Mya *Ty3/Gypsy* peak was only observed in the former. In centromeres of the Nip genome, this most recent *Ty3/Gypsy* peak was not observed, which may indicate that these two types of rice have been diversified for at least 0.77 million years (Figure 5D).

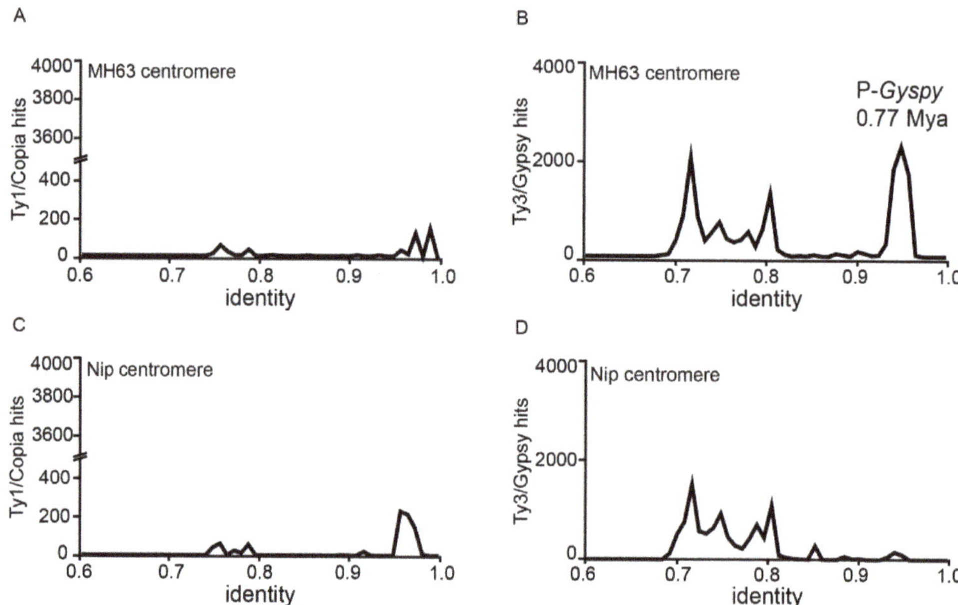

Figure 5. The centromeres of MH63 and Nip were scanned with *Ty1/Copia* and *Ty3/Gypsy* ORF reference sequences and normalized with sequence identities. (**A**) MH63 centromere scanning and normalizing with *ref-M-Ty1/Copia*. (**B–D**) In the same way. Individual peak P-*Gypsy* (*p* value = 0.03) was observed in the MH63 centromeres, and the age of the peaks was 0.77 Mya.

4. Discussion

LTR/TEs have been shown to constitute the major of many monocotyledonous plant genome components, usually, LTR/TEs are randomly inserted across the whole genomes [30]. Earlier inserted TEs may be truncated or fragmented at stochastic sites by various insertional events or sequence mutations that happened later on [31]. Thus, comparisons of the rate in the nucleotide changes between two intact LTR sequences were often used to estimate their insertion time, although it is a challenge to define the exact insertion events for truncated TEs [32]. In our analysis, both rice genomes cumulated plenty of SNPs in the TE ORFs as they were supposed to experience low selection forces during the evolution [33,34]. The huge number of LTR/TE copies, together with their non-conserved nucleotide substitution sites, make them good candidates for evolutionary analysis in several plant systems [2,3,35]. Currently, we have established a matrix-TE approach to comprehensively evaluate the TE insertion events in plant genomes and centromeric regions. This approach overcomes problems related to fragmented TE pieces as well as to stochastic nucleotide substitution sites in the ORFs, as was previously elucidated [3].

Six super-families of LTR/TEs were identified for *Indica* and *Japonica* rice genomes through our super matrix (Figures 1 and 2). The ORFs of intact LTR/TEs were used to generate the identity matrix, then the extracted clusters with high identities were analyzed to figure out the ORF reference sequences by constructing phylogenetic trees (Figure 2). This pipeline has been proven to be an efficient strategy for this type of analysis, because both the intact and truncated TE ORF sequences with diverse SNPs were identified and collected for subsequent evolutionary studies. *Ty1/Copia* and *Ty3/Gypsy* were the two families that showed significant different distribution patterns between the two rice genomes, with that of the other four types showing similar distribution patterns (Figure 3). Detailed analysis showed that significantly more *Ty3/Gypsy* ORFs were inserted in the *Indica* genome, while more *Ty1/Copia* ORFs were inserted in the *Japonica* genome (Table 2). The un-balanced

Ty1/Copia and *Ty3/Gypsy* insertions increased the diversification potential for the *Oryza* genus [5]. These data strongly indicate that *Ty1/Copia* and *Ty3/Gypsy* ORF insertions were probably an important driving force for the differentiation of *Indica* and *Japonica* rice [3,36,37].

LTR/TEs are also major components of the centromeres in both the MH63 and Nip genomes. A significant *Ty3/Gypsy* peak, calculated to appear at around 0.77 Mya, was observed only in MH63 centromeres, suggesting that *Indica* and *Japonica* genomes must be diverged at or before this time point (Figure 5). Previously, by calculating *Ks* values with single-copy ortholog genes, the mean divergence time of *Indica* and *Japonica* rice was at around 0.55 Mya [11]. Since LTR/TE ORFs are generally not under strict evolutionary selection forces as the ortholog genes are, they may accumulate more nucleotide substitutions or other sequence mutations [27,33,34]. We thus conclude that *Indica* and *Japonica* rice may have diverged between 0.55 to 0.77 Mya. For a more accurate estimation of key time points during genome evolution, both ortholog genes as well as LTR/TE ORFs have to be considered simultaneously in a fully balanced way. A recent publication by Zhang et al. successfully applied the GPDF model to interpret the evolution of autopolyploids in the *Saccharum* species [38]. The optimized matrix-TE method may probably be used in other plant species with large genomes.

Supplementary Materials: The following supporting information can be downloaded at: https://www.mdpi.com/article/10.3390/agronomy12071490/s1, Figure S1: The matrix-TE approach pipeline. A rice genome with 12 chromosome sequences was used as the input data and the TE ORFs were used to generate the super matrix. The TE ORF clusters with an identity of over 95% were extracted and annotated. The TE ORF reference sequence was identified by phylogenetic analysis with the whole genomes and the centromeres were scanned and analyzed by the GPDF fit; Table S1: The centromere locations and length statistics for different chromosomes in the MH63 genome; Table S2: The centromere locations and length statistics for different chromosomes in the Nip genome.

Author Contributions: Z.W., W.X., Z.H., Y.W., Y.G. and Y.Z. conceived the bioinformatics experiments and carried out the data analysis. Y.Z. and Z.W. conceived the project and wrote the manuscript. All authors have read and agreed to the published version of the manuscript.

Funding: This work was supported by grants from the Natural Science Foundation of China (No. 21602162, No. 31690090, No. 31690091) and the National Science and Technology Major Project (No. 2016ZX08005003-001).

Data Availability Statement: Not applicable.

Conflicts of Interest: The authors declare no conflict of interest.

References

1. Lisch, D. How important are transposons for plant evolution? *Nat. Rev. Genet.* **2013**, *14*, 49–61. [CrossRef] [PubMed]
2. Lin, J.; Cai, Y.; Huang, G.; Yang, Y.; Li, Y.; Wang, K.; Wu, Z. Analysis of the chromatin binding affinity of retrotransposases reveals novel roles in diploid and tetraploid cotton. *J. Integr. Plant Biol.* **2019**, *61*, 32–44. [CrossRef] [PubMed]
3. Huang, G.; Wu, Z.; Percy, R.G.; Bai, M.; Li, Y.; Frelichowski, J.E.; Hu, J.; Wang, K.; Yu, J.; Zhu, Y. Genome sequence of Gossypium herbaceum and genome updates of Gossypium arboreum and Gossypium hirsutum provide insights into cotton A-genome evolution. *Nat. Genet.* **2020**, *52*, 516–524. [CrossRef] [PubMed]
4. Song, J.M.; Xie, W.Z.; Wang, S.; Guo, Y.X.; Koo, D.H.; Kudrna, D.; Gong, C.; Huang, Y.; Feng, J.W.; Zhang, W.; et al. Two gap-free reference genomes and a global view of the centromere architecture in rice. *Mol. Plant.* **2021**, *14*, 1757–1767. [CrossRef]
5. Li, K.; Jiang, W.; Hui, Y.; Kong, M.; Feng, L.; Gao, L.; Li, P.; Lu, S. Gapless Indica rice genome reveals synergistic contributions of active transposable elements and segmental duplications to rice genome evolution. *Mol. Plant.* **2021**, *14*, 1745–1756. [CrossRef]
6. Möller, M.; Stukenbrock, E.H. Evolution and genome architecture in fungal plant pathogens. *Nat. Rev. Microbiol.* **2017**, *15*, 756–771. [CrossRef]
7. Goff, S.A.; Ricke, D.; Lan, T.H.; Presting, G.; Wang, R.; Dunn, M.; Glazebrook, J.; Sessions, A.; Oeller, P.; Varma, H.; et al. A draft sequence of the rice genome (Oryza sativa L. ssp. japonica). *Science* **2002**, *296*, 92–100. [CrossRef]
8. Yu, J.; Hu, S.; Wang, J.; Wong, G.; Li, S.; Liu, B.; Deng, Y.; Dai, L.; Zhou, Y.; Zhang, X.; et al. A Draft Sequence of the Rice Genome (Oryza sativa L. ssp. indica). *Science* **2002**, *296*, 79–92. [CrossRef]
9. Wang, W.; Mauleon, R.; Hu, Z.; Chebotarov, D.; Tai, S.; Wu, Z.; Li, M.; Zheng, T.; Fuentes, R.R.; Zhang, F.; et al. Genomic variation in 3010 diverse accessions of Asian cultivated rice. *Nature* **2018**, *557*, 43–49. [CrossRef]

10. Huang, X.; Kurata, N.; Wei, X.; Wang, Z.X.; Wang, A.; Zhao, Q.; Zhao, Y.; Liu, K.; Lu, H.; Li, W.; et al. A map of rice genome variation reveals the origin of cultivated rice. *Nature* **2012**, *490*, 497–501. [CrossRef]

11. Stein, J.C.; Yu, Y.; Copetti, D.; Zwickl, D.J.; Zhang, L.; Zhang, C.; Chougule, K.; Gao, D.; Iwata, A.; Goicoechea, J.L.; et al. Genomes of 13 domesticated and wild rice relatives highlight genetic conservation, turnover and innovation across the genus Oryza. *Nat. Genet.* **2018**, *50*, 285–296. [CrossRef] [PubMed]

12. Sun, J.; Ma, D.; Tang, L.; Zhao, M.; Zhang, G.; Wang, W.; Song, J.; Li, X.; Liu, Z.; Zhang, W.; et al. Population Genomic Analysis and De Novo Assembly Reveal the Origin of Weedy Rice as an Evolutionary Game. *Mol. Plant.* **2019**, *12*, 632–647. [CrossRef] [PubMed]

13. Luo, M.C.; Gu, Y.Q.; Puiu, D.; Wang, H.; Twardziok, S.O.; Deal, K.R.; Huo, N.; Zhu, T.; Wang, L.; Wang, Y.; et al. Genome sequence of the progenitor of the wheat D genome Aegilops tauschii. *Nature* **2017**, *551*, 498–502. [CrossRef] [PubMed]

14. Ling, H.Q.; Ma, B.; Shi, X.; Liu, H.; Dong, L.; Sun, H.; Cao, Y.; Gao, Q.; Zheng, S.; Li, Y.; et al. Genome sequence of the progenitor of wheat A subgenome Triticum urartu. *Nature* **2018**, *557*, 424–428. [CrossRef]

15. Springer Nathan, M.; Anderson Sarah, N.; Andorf Carson, M.; Ahern Kevin, R.; Bai, F.; Barad, O.; Barbazuk, W.B.; Bass Hank, W.; Baruch, K.; Ben Zvi, G.; et al. The maize W22 genome provides a foundation for functional genomics and transposon biology. *Nat. Genet.* **2018**, *50*, 1282–1288. [CrossRef]

16. Haberer, G.; Kamal, N.; Bauer, E.; Gundlach, H.; Fischer, I.; Seidel Michael, A.; Spannagl, M.; Marcon, C.; Ruban, A.; Urbany, C.; et al. European maize genomes highlight intraspecies variation in repeat and gene content. *Nat. Genet.* **2020**, *52*, 950–957. [CrossRef]

17. Sun, S.; Zhou, Y.; Chen, J.; Shi, J.; Zhao, H.; Zhao, H.; Song, W.; Zhang, M.; Cui, Y.; Dong, X.; et al. Extensive intraspecific gene order and gene structural variations between Mo17 and other maize genomes. *Nat. Genet.* **2018**, *50*, 1289–1295. [CrossRef]

18. Lippman, Z.; Gendrel, A.V.; Black, M.; Vaughn, M.W.; Dedhia, N.; McCombie, W.R.; Lavine, K.; Mittal, V.; May, B.; Kasschau, K.D.; et al. Role of transposable elements in heterochromatin and epigenetic control. *Nature* **2004**, *430*, 471–476. [CrossRef]

19. Goerner-Potvin, P.; Bourque, G. Computational tools to unmask transposable elements. *Nat. Rev. Genet.* **2018**, *19*, 688–704. [CrossRef]

20. Tang, Y.; Ma, X.; Zhao, S.; Xue, W.; Zheng, X.; Sun, H.; Gu, P.; Zhu, Z.; Sun, C.; Liu, F.; et al. Identification of an active miniature inverted-repeat transposable element mJing in rice. *Plant J.* **2019**, *98*, 639–653. [CrossRef]

21. Zhao, Q.; Feng, Q.; Lu, H.; Li, Y.; Wang, A.; Tian, Q.; Zhan, Q.; Lu, Y.; Zhang, L.; Huang, T.; et al. Pan-genome analysis highlights the extent of genomic variation in cultivated and wild rice. *Nat. Genet.* **2018**, *50*, 278–284. [CrossRef] [PubMed]

22. Deininger, P.; Morales, M.E.; White, T.B.; Baddoo, M.; Hedges, D.J.; Servant, G.; Srivastav, S.; Smither, M.E.; Concha, M.; DeHaro, D.L.; et al. A comprehensive approach to expression of L1 loci. *Nucleic Acids Res.* **2017**, *45*, e31. [CrossRef] [PubMed]

23. El Baidouri, M.; Kim, K.D.; Abernathy, B.; Arikit, S.; Maumus, F.; Panaud, O.; Meyers, B.C.; Jackson, S.A. A new approach for annotation of transposable elements using small RNA mapping. *Nucleic Acids Res.* **2015**, *43*, e84. [CrossRef]

24. Jiang, N.; Bao, Z.; Zhang, X.; Hirochika, H.; Eddy, S.R.; McCouch, S.R.; Wessler, S.R. An active DNA transposon family in rice. *Nature* **2003**, *421*, 163–167. [CrossRef]

25. Chen, J.; Lu, L.; Benjamin, J.; Diaz, S.; Hancock, C.N.; Stajich, J.E.; Wessler, S.R. Tracking the origin of two genetic components associated with transposable element bursts in domesticated rice. *Nat. Commun.* **2019**, *10*, 641. [CrossRef] [PubMed]

26. Carpentier, M.C.; Manfroi, E.; Wei, F.J.; Wu, H.P.; Lasserre, E.; Llauro, C.; Debladis, E.; Akakpo, R.; Hsing, Y.I.; Panaud, O. Retrotranspositional landscape of Asian rice revealed by 3000 genomes. *Nat. Commun.* **2019**, *10*, 24. [CrossRef]

27. Liao, Y.; Zhang, X.; Li, B.; Liu, T.; Chen, J.; Bai, Z.; Wang, M.; Shi, J.; Walling, J.G.; Wing, R.A.; et al. Comparison of Oryza sativa and Oryza brachyantha genomes reveals selection-driven gene escape from the centromeric regions. *Plant Cell* **2018**, *30*, 1729–1744. [CrossRef]

28. Ma, J.; Bennetzen, J.L. Rapid recent growth and divergence of rice nuclear genomes. *Proc. Nati. Acad. Sci. USA* **2004**, *101*, 12404–12410. [CrossRef]

29. Chuong, E.B.; Elde, N.C.; Feschotte, C. Regulatory activities of transposable elements: From conflicts to benefits. *Nat. Rev. Genet.* **2017**, *18*, 71–86. [CrossRef]

30. Jain, M.; Nijhawan, A.; Tyagi, A.K.; Khurana, J.P. Validation of housekeeping genes as internal control for studying gene expression in rice by quantitative real-time PCR. *Biochem. Biophys. Res. Commun.* **2006**, *345*, 646–651. [CrossRef]

31. Meyers, B.C.; Tingey, S.V.; Morgante, M. Abundance, distribution, and transcriptional activity of repetitive elements in the maize genome. *Genome Res.* **2001**, *11*, 1660–1676. [CrossRef] [PubMed]

32. Sultana, T.; Zamborlini, A.; Cristofari, G.; Lesage, P. Integration site selection by retroviruses and transposable elements in eukaryotes. *Nat. Rev. Genet.* **2017**, *18*, 292–308. [CrossRef] [PubMed]

33. SanMiguel, P.; Gaut, B.S.; Tikhonov, A.; Nakajima, Y.; Bennetzen, J.L. The paleontology of intergene retrotransposons of maize. *Nat. Genet.* **1998**, *20*, 43–45. [CrossRef] [PubMed]

34. Biémont, C.; Vieira, C. Genetics: Junk DNA as an evolutionary force. *Nature* **2006**, *443*, 521–524. [CrossRef]

35. Feschotte, C.; Jiang, N.; Wessler, S.R. Plant transposable elements: Where genetics meets genomics. *Nat. Rev. Genet.* **2002**, *3*, 329–341. [CrossRef]

36. Huang, G.; Huang, J.Q.; Chen, X.; Zhu, Y.X. Recent advances and future perspectives in cotton research. *Annu. Rev. Plant. Biol.* **2021**, *72*, 437–462. [CrossRef]

37. Wang, K.; Huang, G.; Zhu, Y. Transposable elements play an important role during cotton genome evolution and fiber cell development. *Sci. China Life Sci.* **2016**, *59*, 112–121. [CrossRef]
38. Zhang, Q.; Qi, Y.; Pan, H.; Tang, H.; Wang, G.; Hua, X.; Wang, Y.; Lin, L.; Li, Z.; Li, Y.; et al. Genomic insights into the recent chromosome reduction of autopolyploid sugarcane *Saccharum spontaneum*. *Nat. Genet.* **2022**, *54*, 885–896. [CrossRef]

Article

Genome-Wide Analysis of the Rice Gibberellin Dioxygenases Family Genes

Yurong He, Wei Liu, Zhihao Huang, Jishuai Huang, Yanghong Xu, Qiannan Zhang and Jun Hu *

State Key Laboratory of Hybrid Rice, Engineering Research Center for Plant Biotechnology and Germplasm Utilization of Ministry of Education, College of Life Sciences, Wuhan University, Wuhan 430072, China; heyurong@whu.edu.cn (Y.H.); liuwei233333@163.com (W.L.); 2016301060003@whu.edu.cn (Z.H.); huangjishuai28@whu.edu.cn (J.H.); 13100699733@163.com (Y.X.); qiannanzhang813@163.com (Q.Z.)
* Correspondence: junhu@whu.edu.cn

Abstract: Gibberellins (GAs), a pivotal plant hormone, play fundamental roles in plant development, growth, and stress response. In rice, *gibberellin-dioxygenases* (*GAoxes*) are involved in the biosynthesis and deactivation of gibberellins. However, a comprehensive genome-wide analysis of GA oxidases in rice was not uncovered. Here, a total of 80 candidate *OsGAox* genes were identified and 19 *OsGAox* genes were further analyzed. Studies on those 19 *OsGAox* genes, including phylogenetic tree construction, analysis of gene structure, exploration of conserved motifs and expression patterns, were conducted. Results showed that the *GAox* genes in *Arabidopsis* and rice were divided into four subgroups and shared some common features. Analysis of gene structure and conserved motifs revealed that splicing phase and motifs were well conserved during the evolution of *GAox* genes in *Arabidopsis* and rice, but some special conserved motifs possessed unknown functions need to be further studied. Exploration of expression profiles from RNA-seq data indicated that each *GAox* gene had tissue-specific expression patterns, although they varied greatly. The expression patterns of these genes under GA_3 treatment revealed that some genes, such as *OsGA2ox1*, *OsGA2ox3*, *OsGA2ox4*, *OsGA2ox7*, *OsGA20ox1*, and *OsGA20ox4*, may play a major role in regulating the level of bioactive GA. Taken together, our study provides a comprehensive analysis of the *GAox* gene family and will facilitate further studies on their roles in rice growth and development so that these genes can be better exploited.

Keywords: genome-wide analysis; gibberellin; *GAox*; rice (*Oryza sativa*)

Citation: He, Y.; Liu, W.; Huang, Z.; Huang, J.; Xu, Y.; Zhang, Q.; Hu, J. Genome-Wide Analysis of the Rice Gibberellin Dioxygenases Family Genes. *Agronomy* **2022**, *12*, 1627. https://doi.org/10.3390/ agronomy12071627

Academic Editor: Matthew Hegarty

Received: 4 June 2022
Accepted: 27 June 2022
Published: 7 July 2022

Publisher's Note: MDPI stays neutral with regard to jurisdictional claims in published maps and institutional affiliations.

1. Introduction

Bioactive gibberellins (GAs), a large group of diterpene plant hormones, play essential roles in the complete life cycle of rice, such as seed germination, stem elongation, root development, pollen development, and flower induction [1–5]. To date, about 136 GAs have been identified in higher plants, fungi, and bacteria [6], while most of them are non-bioactive and act as precursors for the bioactive forms or deactivated metabolites and only GA_1, GA_3, GA_4, and GA_7 are bioactive GAs [7]. In higher plants, the GA metabolism pathway has been extensively studied [8]. This pathway mainly involves three stages of reactions, according to the location and the enzymes involved. In the first stage, geranylgeranyl diphosphate (GGDP), a common C20 precursor for diterpenoids, is converted to the tetracyclic hydrocarbon intermediate ent-kaurene by two kinds of terpene synthases (TPSs) in the plastids, ent-copalyl diphosphate synthase (CPS) and ent-kaurene synthase (KS). In the second stage, GA_{12} and GA_{53} are synthesized from ent-kaurene by two types of cytochrome P450 monooxygenases (P450s) at the endoplasmic reticulum, ent-Kaurene oxidase (KO), and entkaurenoic acid oxidase (KAO). In the final stage of the pathway, bioactive GA synthesis is catalyzed by two kinds of soluble 2-oxoglutarate-dependent dioxygenases (2ODDs) known as GA 20-oxidase (GA20ox) and GA 3-oxidase (GA3ox) in the cytosol and the bioactive GAs or their immediate precursors are inactivated by a third 2ODD, GA 2-oxidase (GA2ox), including C19-GA2oxes and C20-GA2oxes [9].

The GA20ox, GA3ox, and GA2ox belong to the 2-ODDs superfamily and are each encoded by a multigene family [10], the members of which have different expression patterns and thus regulate GA metabolism in different plant developmental processes. Most *GAox* gene family members have been cloned and identified and their biological functions have also been studied in various plant species [11,12]. By manipulating the expression of *GAox* genes, the levels of endogenous active GAs can be regulated. For example, the deficiency of a rice semidwarfing gene (*sd-1/OsGA20ox2*) known as the "green revolution gene" causes reduction in endogenous GAs, thus affecting the plant height in rice [13]. Therefore, it is extremely important to identify and exploit *GAox* genes in plants.

Rice (*Oryza sativa* L.), one of the most important global food crops, is a primary source of food for over half of the world's population [14]. Historically, the first green revolution resulted from the utilization of *GAox* genes and fertilizer greatly increased the production of crops, especially rice and wheat [3]. However, considering the deterioration of the environment caused by the use of fertilizer and the increasing of population, rice production should be increased more quickly via more scientific strategies. Grain yield in rice is a complex trait affected by multiple factors and major progress in increasing rice yield is on the basis of the exploitation of high yield varieties. Due to the key role that GA plays in regulating the yield of the rice, a lot of high yield varieties have been exploited by regulating genes involved in the GA biosynthetic pathways in rice [15–17]. In order to better manipulate the *GAox* genes to attain high yield varieties, further study of the *GAox* family genes needs to be conducted.

Here, we conducted a comprehensive genome-wide analysis of *GAox* genes in rice and further studied 19 *OsGAox* genes, including analysis of the phylogenetic relationship, gene structure, and motif identification. Furthermore, we investigated the gene expression profiles in different rice organs and alteration under GA_3 treatment. These data suggested *OsGAox* genes had special expression patterns in various organs and played different roles under GA_3 treatment in rice. This study will serve as a foundation into a comprehensive and deep study of *GAox* genes in future research.

2. Materials and Methods

2.1. Plant Materials and Treatments

Rice (*Oryza sativa* L. japonica cv. Nipponbare) seeds were grown in sterile water in conical flasks at 37 °C in an incubator. After 3 days, the seeds were germinated and transferred into hydroponic boxes. Seedlings were cultivated in greenhouse, and watered with nutrient solution every day; 7 days later, some of the seedlings were transferred into a paddy at Wuhan University under natural conditions, while the rest of them were still cultivated in greenhouse. The temperature of the greenhouse was 27 °C, and the relative humidity was 80%. Plant materials for analyzing expression patterns in different periods included: (i) stigma, ovary and root; (ii) young leaves, mature leaves; (iii) different developmental stages of panicles and seeds, which were sampled according to the rice growth stages in the paddy. For the GA_3 treatment, the 14-day seedlings of rice cultivated in greenhouse were treated with stress condition: 100 μM GA_3 (CAS:77-06-5, Sigma-Aldrich, St. Louis, MO, USA), which was dissolved in anhydrous ethanol and mixed with nutrient solution. For the Mock treatment, the 14-day seedlings of rice cultivated in greenhouse were still cultured with nutrient solution. At the same time as the GA_3 treatment, the Mock treatment was also carried out, and samples were taken at the same time. During the time period of GA_3 and Mock treatment, the leaves were collected at 0, 2, 4, 6, 8, 10, and 12 h after treatment and quickly frozen in liquid nitrogen for RNA extraction.

2.2. Sequence Retrieval and Identification and Analysis of GAox Genes in Rice

All sequences of 21 *GAox* genes were downloaded from Phytozome (The Plant Comparative Genomics portal of the Department of Energy's Joint Genome Institute) [18], which was described in a previous report [17]. Finally, SMART (available online: http://smart.emblheidelberg.de/ (accessed on 15 October 2020)) were employed to further confirm the existence of both 2OG-FeII_Oxy (PF03171) and DIOX_N (PF14226) domains in all identified *GAox* gene proteins. All *GAox* genes' loci and their information, including chromosome location, were searched in the RGAP and DAP-DB (the Rice Annotation Project Database). The map of genes distributed across chromosomes was created with Mapchart software.

2.3. Analysis of Phylogenetic Relationship and Sequences

Multiple sequence alignments of 19 OsGAox proteins from rice and 16 AtGAox proteins from *Arabidopsis* were performed by using the iTOL website [19]. A phylogenetic tree was conducted using the neighbor-joining (NJ) method with 1000 bootstrap replicates in TreeBeST (available online: https://github.com/Ensembl/treebest (accessed on 22 September 2021)) [20]. For gene structure analysis, the genomic DNA sequences with the corresponding cDNA sequences from the RADP and RAP-DB database, including the exons and introns structures of individual *GAox* genes, were displayed via the Gene Structure Display Server (available online: http://gsds.cbi.pku.edu.cn/ (accessed on 7 May 2021)) [21]. Conserved motifs analysis was performed by MEME analysis tool with the maximum number of motifs to identify set to 10 [22].

2.4. Analysis of GA Oxidase Family Gene Expression Patterns

To study expression patterns of *OsGAox* genes, the public RNA-seq data contained a wide range of rice developmental stages, downloaded from Rice Functional Genomic Express Database (available online: http://signal.salk.edu/cgi-bin/RiceGE (accessed on 23 March 2021)). Subsequently, the heat map was constructed with Heml software with the log-transformed values by reanalyzing the RNA-seq data [23].

2.5. RNA Extraction, qRT-PCR and RNA seq

Total RNA was isolated from collected samples using TRIzol reagent (Takara, Japan) and then treated with DNase I (New England Biolabs, Beijing, China) according to the manufacturers' instructions. The RNA concentration was measured using RNA Nano6000 Assay Kit of the Bioanalyzer 2000 system (Thermo Scientific, Thermo Fisher Scientific Inc.1.6.198, Waltham, MA, USA). Approximate 1 µg of total RNA were reverse-transcribed using ABScript III RT Master Mix for qPCR with gDNA Remover (ABclonal; cat.no.RK20429). Reverse transcription polymerase chain reaction (RT-PCR) was performed using the Hieff qPCR SYBR Green Master Mix (YEASEN; cat.no.11201ES08) in a CFX96 Real-Time System (BIO-RAD, CFX Manager Software 3.1, Hercules, CA 94547, USA). Specific primers (Supplementary Table S3) for qRT-PCR were designed by using the software Primer Premier 5 and synthesized by TSINGKE (Wuhan, China), and were then checked with the cDNA by PCR. The amplification length for each gene was restricted to 80–250 bp to ensure the efficiency of optimal polymerization. Quantitative real-time PCR (qRT-PCR) was performed with 1 µL cDNA, 0.4 µM gene specific primers, and 10 µL 2× mix (YEASEN), and water was used to supplement to 20 µL by a CFX96 Real-Time System (BIO-RAD, CFX Manager Software 3.1, Hercules, CA 94547, USA) according to the manufacturer's instruction. The reaction program of qRT-PCR was performed under the following conditions: 95 °C for 5 min, followed by 40 cycles at 95 °C for 10 s, and 60 °C for 30 s. Three replicates were carried out for each sample and three biological replicates were also performed for each sample. The relative gene expression levels were calculated using a $2^{-\Delta\Delta Ct}$ method [24] and the melting curve was carried out for each PCR product to avoid nonspecific amplification. The rice gene *actin* (*LOC_Os03g50885*) was used as an internal control to normalize the expression of related genes involved in GAs biosynthesis.

2.6. Statistical Analysis

All data were analyzed using the GraphPad Prism 7.00 statistics program (https://www.graphpad.com/, accessed on 23 March 2021) and the means were compared by Student's *t*-test. Each assay was performed in three biological replicates and technical replications.

3. Results

3.1. Identification and Analysis of the GAox Family Genes in the Rice

With genome wide analysis, we identified a total of 80 candidate *OsGAox* genes based on genome and transcriptome databases (Supplementary Table S1). A previous study reported about 21 *OsGAox* genes in rice [25]. Based on the results of alignment, two genes (*OsGA2ox2* and *OsGA2ox10*) lacked a common domain DIOX_N (PF14226) were excluded in this study. To better understand the distribution of rice *GAox* genes on chromosomes, the map of genes distributed across chromosomes was created with Mapchart software (Figure 1). Our study showed that the 19 *OsGAox* genes were unevenly distributed on seven chromosomes. There were four *OsGAox* genes on both chromosome 1 and chromosome 4. Five *OsGAox* genes were located on chromosome 5. Two *OsGAox* genes were mapped on both chromosome 2 and chromosome 7. Only one *OsGAox* gene was found on chromosome 2 and chromosome 8. Interestingly, this result showed undiscovered GA biosynthesis genes in the other five chromosomes. The gene names, entry ID, number of deduced amino acid, molecular weights, predicted subcellular localizations, group classifications, and theoretical pI were also summarized in Table 1. Data showed that the protein length of the identified GAox family genes ranged from 301 (OsGA20ox6) to 446 (OsGA20ox4) amino acids (aa), with an average of 363 aa. The molecular mass ranged from 32.10 to 47.63 kDa, and the pI ranged from 5.25 (OsGA20ox8) to 7.44 (OsGA2ox6). Most of GAox family genes were predicted to be located in the nucleus and cytoplasm on WoLFPSORT [26] and TargetP [27], which was consistent with the previous findings [15,28]. Results also suggested some of them can be transported into mitochondria or chloroplasts, implied that these organelles might be also involved in the GA metabolism in plants.

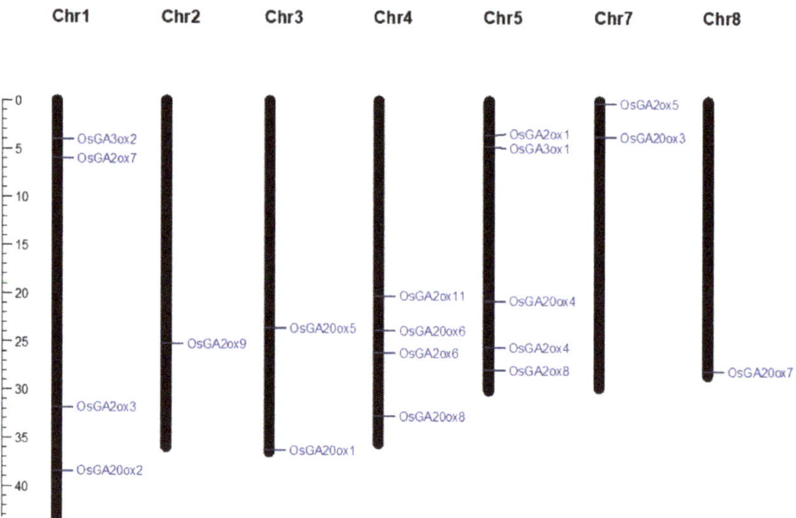

Figure 1. The distribution of 19 *GAox* genes on rice chromosomes. The scale bar on the left represented the length (Mb) of rice chromosomes.

Table 1. The 19 OsGAox family genes in the rice.

Gene Names	Entry ID	Number of Deduced Amino Acid	Molecular Weight(kDa)	Subcellular Location (TargetP 2.0)	Subcellular Location (WoLFPSORT)	Groups	Theoretical pI
OsGA2ox1	Os05g06670	383	40.62	other/sp	pero/cyto/nucl	C19GA2ox	7.11
OsGA2ox3	Os01g55240	328	35.33	other/sp	cyto/chlo/pero	C20GA2ox	6.67
OsGA2ox4	Os05g43880	355	37.79	other/sp	chlo/cyto/nucl	C19GA2ox	6.86
OsGA2ox5	Os07g01340	342	38.69	other	nucl/cyto/chlo	C19GA2ox	6.27
OsGA2ox6	Os04g44150	359	39.04	other/sp	cyto/chlo/pero	C20GA2ox	7.44
OsGA2ox7	Os01g11150	336	35.14	other/sp/mito/chlo	chlo/plas/cyto	C19GA2ox	7.08
OsGA2ox8	Os05g48700	354	37.31	other/chlo	cyto/chlo/pero	C19GA2ox	6.4
OsGA2ox9	Os02g41954	360	39.02	other/sp	cyto/nucl/mito	C20GA2ox	5.66
OsGA2ox11	Os04g33360	372	40.23	other/mito	chlo/nucl/cyto	C20GA2ox	6.71
OsGA3ox1	Os05g08540	385	41.56	other/chlo	chlo/mito	GA3ox	6.36
OsGA3ox2	Os01g08220	374	40.57	other/sp	cyto/chlo/nucl	GA3ox	6.96
OsGA20ox1	Os03g63970	373	42.26	other/sp	cyto/chlo/nucl	GA20ox	6.41
OsGA20ox2	Os01g66100	390	42.51	other/sp	cyto/nucl/plas	GA20ox	6.01
OsGA20ox3	Os07g07420	368	40.49	other/sp	chlo/plas	GA20ox	6.09
OsGA20ox4	Os05g34854	446	47.63	other/chlo	chlo/cyto/nucl	GA20ox	7.18
OsGA20ox5	Os03g42130	353	39.29	other/sp	cyto/chlo/mito	GA20ox	4.98
OsGA20ox6	Os04g39980	301	32.1	other/sp	cyto/mito	GA20ox	5.61
OsGA20ox7	Os08g44590	384	41.83	other/sp	cyto	GA20ox	6.35
OsGA20ox8	Os04g55070	327	35.79	other/sp	cyto/chlo/nucl	GA20ox	5.25

Note: WoLFPSORT an d TargetP 2.0 were used to predict the subcellular localization of the 19 OsGAox family genes; pero: peroxisome; nucl: nucleus; chlo: chloroplast;cyto: cytoplasmic; mito: mitochondrion; plas: plasma membrane; sp: signal peptide. Groups were classified based on phylogenetic trees.

3.2. Phylogenetic Analysis of the GAox Gene Family

In *Arabidopsis*, sixteen *GAox* genes (seven *GA2oxes*, four *GA3oxes* and five *GA20oxes*) have been identified [9]. In order to determine evolutionary relationships of rice and *Arabidopsis GAoxes*, the phylogenetic tree was constructed using the neighbor-joining (NJ) method from alignments of the complete protein sequences from 16 AtGAox and 19 OsGAox (Figure 2). The phylogenetic tree generated four distinct subgroups, and also revealed that the phylogenetic representation of *Arabidopsis* and rice GAox proteins was diverse. Among the 35 proteins, five OsGAox and five AtGAox belonged to C19-GA2ox subfamily, eight OsGAox and five AtGAox belonged to GA20ox subfamily, two OsGAox and four AtGAox belonged to GA3ox subfamily, and four OsGAox and two AtGAox belonged to C20GA2ox subfamily. Four subfamilies (GA20ox, GA3ox, C19GA2ox, and C20GA2ox) were shared in both two species, suggesting that these four subfamilies might be widespread in plant GA metabolism. Furthermore, the numbers of *GA20ox* and *GA2ox* genes were greater than *GA3ox* in both two species, indicating that *GA20ox* and *GA2ox* had undergone a more dynamic evolutionary route than *GA3ox* and thus resulted in more functional redundancy. Overall, the *GAox* genes shared some common characteristics in monocots' and dicots' evolutionary relationship, so the related studies on them could interact and put each other forward.

3.3. Gene Structure and Conserved Motif Analysis of GAox Gene Family

To support the phylogenetic analysis, we performed gene structure analysis of *GAox* family members from *Arabidopsis* and rice. As shown in Figure 3, the the number of exons was conserved, ranging from 1 to 3 exons in *AtGAox* and *OsGAox* genes. We also investigated intron phases with respect to codons. Most of the first intron was a phase 2 intron which meant that splicing occurred after the second nucleotide and the second intron is generally a phase 0 intron, which meant that splicing occurred after the third nucleotide. This result revealed that the splicing phase was also well conserved during the evolution of *GAox* genes in *Arabidopsis* and rice. Generally, the distribution is unequal in intron phases with a bias in favor of phase 0, which indicated that the ancient introns were dominantly of phase 0, so as to favor intron average length of *GA oxidase* genes in rice shuffling [29–31]. In order to better understand the structure of the *GAox* gene family, we further analyzed the cis-acting elements contained in the first 2000 bp of the

19 *OsGAox* genes (Figure S1). Most genes have roughly the same cis-acting elements, but different cis-acting elements are distributed in different positions of each gene, which may provide help for studying how to regulate *GAox* gene expression. To further investigate the relationship among GAox proteins with the same subfamily in rice, 10 different conserved motifs were identified and examined by using the MEME motif search tool in rice (Figures S2 and 4). Motifs 1, 2, 3, 4, 5, 6, and 8 were shared by most of the GAox proteins. Interestingly, there were some specific absent motifs in specific subfamilies. Motif 7 was absent in all the members of C19-GA2ox except for GA2ox8 and motif 10 was shared by all C20-GA2ox and GA3ox. Motif 9 was only shared by proteins of the C19-GA2ox subfamily, which revealed that motif 9 may have a special function in C19-GA2ox subfamily. Differences among motif distributions explained sources of functional divergence in GA oxidases in evolutionary history. However, the function of these motifs in these proteins needed to be investigated further.

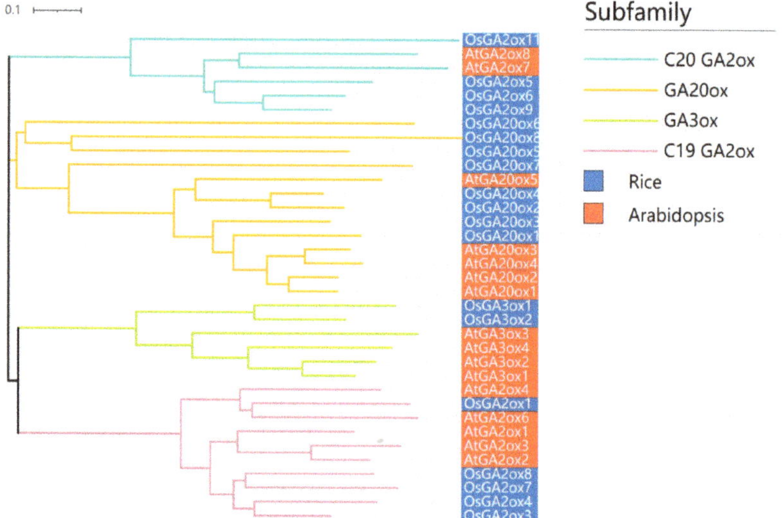

Figure 2. Phylogenetic relationships in *AtGAox* and *OsGAox* genes. The phylogenetic tree was constructed based on the complete protein sequences of 16 AtGAox and 19 OsGAox proteins using MEGA 7.0 software. *Arabidopsis* and rice GAoxs were labeled by red and blue, respectively.

3.4. Expression Patterns of GAox Genes

To provide clues for functional studies of *GAox* genes, we used FPKM values to represent their expression profiles in different organs of rice in this study [32]. Due to the lack of the corresponding probe of *OsGA3ox2*, we analyzed expression patterns of the other 18 genes (Supplementary Table S2). Our results showed that the *OsGAox* genes had different expression levels in various organs (Figure 5). It was worthy to note that some *OsGAox* genes, especially *OsGA2ox3*, *OsGA2ox7*, *OsGA2ox8*, and *OsGA20ox6*, were highly expressed in panicle and *OsGA2ox7* and *OsGA20ox6* had a high expression in all organs; however, *OsGA20ox8* had a low expression in all organs. Based on the above results, we verified the expression pattern of these five genes from RNA-seq data by qRT-PCR. As shown the Supplementary Figure S3, those genes basically had the same pattern compared to RNA-seq results. Overall, those results represented that each *OsGAox* gene possessed special expression pattern in various organs, thus promoting the study on potential functions of these genes in different developmental stages of rice.

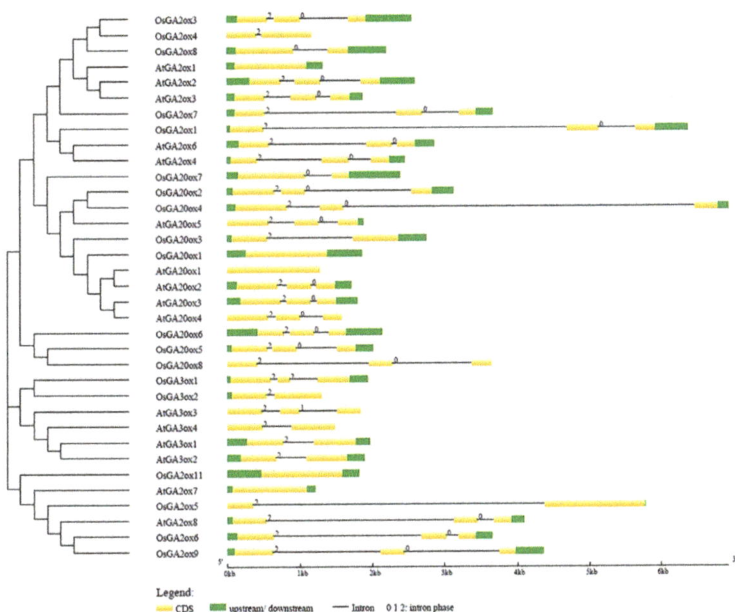

Figure 3. Exon-intron structure of *AtGAoxes* and *OsGAoxes*. Exons and introns are shown by filled yellow boxes and thin black lines, respectively. UTRs are displayed by filled green boxes. Intron phases 0, 1, and 2 are indicated by numbers 0, 1, and 2.

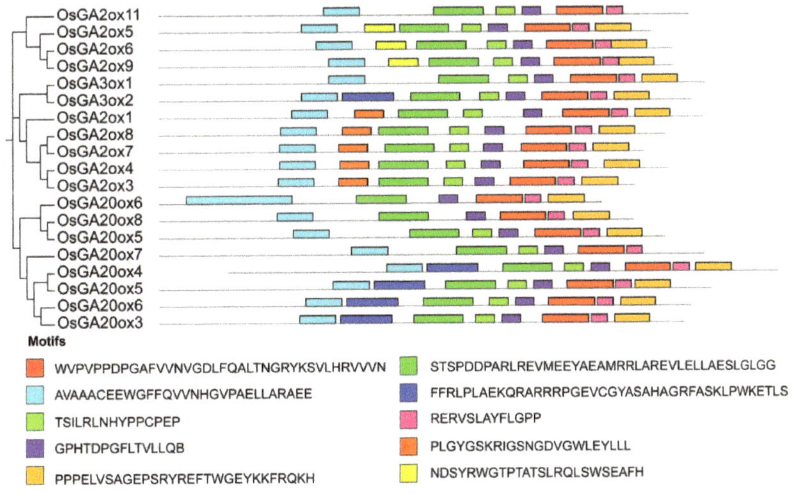

Figure 4. Motif identification of OsGAox proteins. Different motifs are marked in numbers 1–10. The sequence logos and E values for each motif are given in Figure S2.

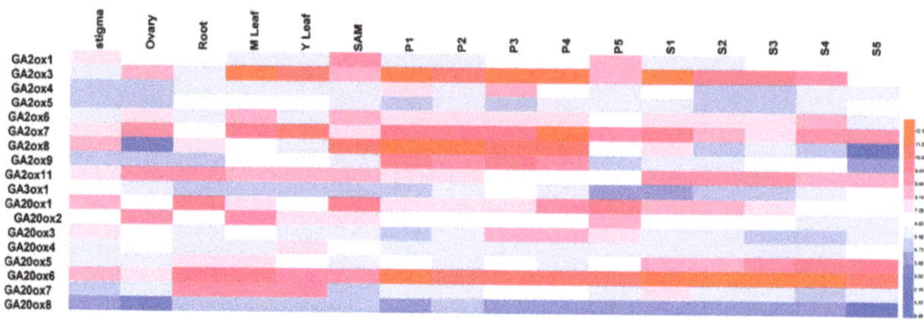

Figure 5. Tissue-specific expression of *OsGAoxes* in various tissues. The expression values were normalized with logarithm with the base of 2 using the HemI software. The color bar in heat maps represents the expression values: blue represents low expression, white represents no significant difference in expression, and red means high expression. M Leaf: mature leaf; Y Leaf: young leaf; SAM means shoot apical meristem; P1 means young inflorescence of up to 3 cm; P2 means Inflorescence of 3–5 cm; P3 means Inflorescence of 5–10 cm; P4 means Inflorescence of 10–15 cm; P5 means 15–22 cm; S1 means seed of 0–2 DAP (Days After Pollination); S2 means seed of 3–4 DAP; S3 means seed of 5–10 DAP; S4 means seed of 11–20 DAP; S5 means seed of 21–29 DAP.

3.5. GAox Gene Expression Profiles under GA3 Treatment

To explore the manner of these 19 *OsGAox* genes in response to exogenous GA, we treated the 14-day rice seedlings with GA_3, then took the leaves of rice seedlings in different time periods and analyzed the expression profiles of 19 *OsGAox* genes under GA_3 treatment by qRT-PCR. As expected, most of them exhibited different expression alteration under GA_3 treatment (Figure 6). Results showed that the expression levels of *OsGA2ox1*, *OsGA2ox3*, *OsGA2ox4*, *OsGA2ox7*, *OsGA20ox1*, and *OsGA20ox4* exhibited an upward trend after GA3 treatment, while the expression levels of *OsGA2ox5*, *OsGA2ox8*, *OsGA3ox1*, *OsGA20ox2*, and *OsGA20ox6* increased first and then decreased. Furthermore, while the expression level of *OsGA3ox2* was suppressed, the expression of *OsGA2ox9* and *OsGA2ox11* had no significant change. For the remaining GAox gene family members, because their Ct value is greater than 35, the test results cannot quantitatively analyze the gene expression, so they are not released. Overall, those results indicated that not all *OsGAox* genes were involved in the gibberellin homeostasis; some *OsGAox* genes are the main regulatory genes, but some genes which were not involved in the gibberellin homeostasis may play a role in other aspects of rice developmental progress.

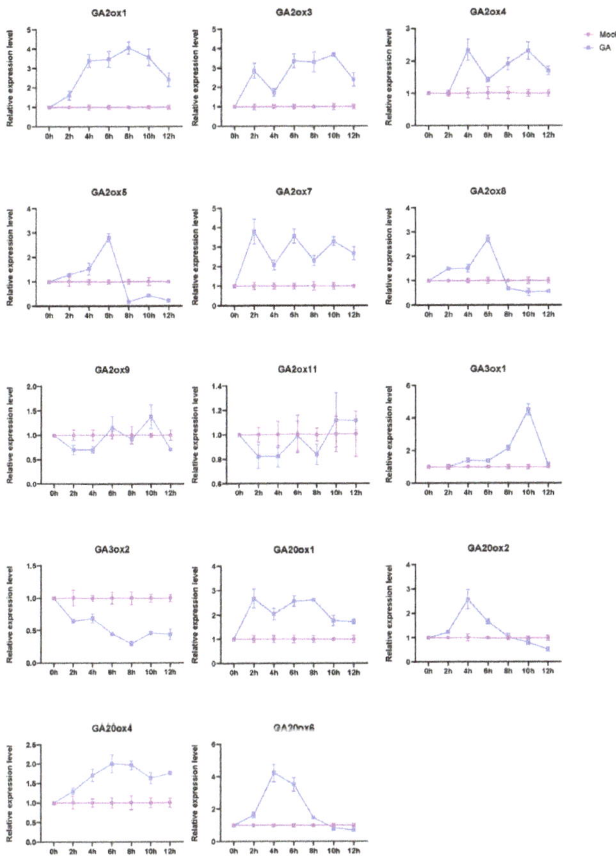

Figure 6. The expression profiles of 19 *OsGAox* genes under GA$_3$ treatment in 14-day rice seedlings. Mock represents controls (non-treated plants), GA represents GA$_3$ treatment. The relative gene expression levels were calculated using the $2^{-\Delta\Delta Ct}$ method and expressed as the fold change relative to expression of the Mock.

4. Discussion

The 2-ODDs superfamily is a large family that has been identified in many land plants, especially in crops, such as cucumber, soybean, and rice [25,33]. Previous research showed that 21 *OsGAox* genes have been named in rice [25], but most of researchers majorly focused on the cloning, function analysis, and molecular mechanism of one certain *OsGAox* gene in rice. Here, we identified 80 *OsGAox* genes in the rice genome and further analyzed 19 *OsGAox* genes with both 2OG-FeII_Oxy (PF03171) and DIOX_N(PF14226) domains for summarizing their common characteristics. The studies included phylogenetic tree construction, analysis of gene structure and conserved motifs, investigation of their expression patterns and exploration of their function of regulating the level of bioactive GA.

In our evolutionary analysis, GAox proteins family in *Arabidopsis* and rice can be divided into four subfamilies based on their protein sequences. Phylogenetic analysis revealed distinct differences between the two species, such as the number of each subfamily (Figure 2). In some studies, there was another subfamily in some species, *GA7ox* [33–35]. GA7ox, which oxidizes GA12-aldehyde to GA12 and possesses mono-oxygenase 7-oxidase activity, was reported in pumpkin and cucumber but has not been found in other species. So far, although three *GA2ox, GA3ox,* and *GA20ox* subfamilies have been found in most plant species, the identification of the *GAox* gene in other plant species still needs to be performed for finding *GA7ox* or the other subfamily so that we can have a more comprehensive understanding of the *GAox* genes' evolutionary relationship.

In the present study, gene structure revealed that *OsGAox* and *AtGAox* gene structures were conserved and the ancient introns in rice were dominantly of phase 0 to favor intron average length of *GAox* genes in rice shuffling. Conserved motif analysis of the GAox proteins in rice revealed that most of motifs consisted in all GAox, while a few motifs were only possessed by a certain subfamily. To better understand the function of *GAox* genes, those special motifs' biological functions are waiting for further investigating and exploring.

In this study, the expression patterns of 19 *OsGAox* genes were observed in different rice organs. The result showed that expression levels of these genes varied greatly in different organs. Four genes, *OsGA2ox3*, *OsGA2ox7*, *OsGA2ox8*, and *OsGA20ox6*, were highly expressed in panicle, indicating that they may play a key role in panicle development. *OsGA20ox6* and *OsGA20ox8* had a higher and lower expression level in all organs, respectively, revealing that they may have different roles in all developmental processes of rice. Gene expression patterns can generate fundamental new insights into their biological function in organisms in general. The *OsGA20ox6* gene, for example, is essential for reproductive development, including anther dehiscence, pollen fertility, and seed initiation in rice [36], which is consistent with its expression patterns in various organs. From these results, the different expression levels in all rice organs indicated that certain *OsGAox* genes might play important roles in plant development and have unique functions in specific developmental stages. Now many researchers are cloning new *GAox* genes and exploring their functions, so our results can facilitate their research.

The levels of bioactive GAs in plants are maintained via feedback and feedforward regulation of GA metabolism [37], so almost all *OsGAox* genes expression level changed in response to GA_3 and Mock treatment. We also examined *OsGAox* gene expression patterns in rice seedling under GA_3 and Mock treatment. Notably, most *OsGAox* genes showed distinct changes after treatment. The expression levels of some genes were increased after GA_3 treatment, but some of them were increased first and then decreased, which indicated that there might be feedback regulation to inhibit the expression of these genes after GA_3 treatment for a period of time. *OsGA3ox2* has been inhibited, which indicates that it may play a negative regulatory role in the GA synthesis pathway. Nevertheless, *OsGA2ox9* and *OsGA2ox11* have no significant change, which indicates that they may not participate in the response to exogenous GA. However, the other genes presented unexpected expression patterns which were inconsistent with our outstanding of the GA feedback and feedforward regulation (Figure 6). We speculated that some genes in the rice *GAox* gene family may maintain linkage relationships during the evolution process, and this linkage may have some advantages in selection under certain conditions, so we further analyzed the collinearity of these 19 *OsGAox* genes (Figure S4), which provided useful clues for studying the relationship between gene functions. GAs regulate plant growth and development and are closely related to the yield of other crops [1–5], and GA oxidases play pivotal roles in GAs' biosynthesis and metabolism [7], but how to coordinate the regulation mechanism between the two has not yet been clear. Therefore, how all the *GAox* genes combine and coordinate to regulate the bioactive GA level should be further studied in the future. In addition, considering that the feasibility of genetic improvement of rice yield by manipulating the expression of *OsGAox* genes [38], our results can be for reference to improve the rice yield by changing one certain *OsGAox* gene expression.

5. Conclusions

In this study, we comprehensively analyzed 19 *OsGAox* genes in rice, which can be divided into four subgroups according to phylogenetic tree. Gene structure and conserved motif analysis showed that most *GAox* genes were conserved in two model plants, *Arabidopsis* and rice. We also analyzed their expression profiles in different organs in rice. The result suggested that various *OsGAox* genes played different roles in rice developmental stages. In addition, the expression patterns of those genes under GA$_3$ treatment were also explored and from the result we found out not all *OsGAox* genes were involved in the gibberellin homeostasis. Taken together, our data will generate insight into the further study of *OsGAox* genes in rice and provide reference for exploitation of certain *GAox* genes to improve the rice yield and food security.

Supplementary Materials: The following supporting information can be downloaded at: https://www.mdpi.com/article/10.3390/agronomy12071627/s1. Figure S1. The cis-acting elements contained in the −2000 bp of the 19 *OsGAox* genes. Figure S2. The sequence logos and E values for 10 conserved motifs among OsGAox proteins. Figure S3. Validation of expression patterns of 5 *OsGAox* genes by qRT-PCT. Figure S4. The collinearity analysis of the 19 *OsGAox* genes in rice. Table S1. 80 candidate *OsGAox* genes. Table S2. FPKM values of *OsGAox* genes in rice different organs. Table S3. Primers used in this study.

Author Contributions: J.H. (Jun Hu) designed the research; Y.H., W.L., Z.H., Q.Z., J.H. (Jishuai Huang) and Y.X. performed the molecular biology experiments; Z.H. performed the bioinformatics analyses; Y.H., W.L. and J.H. (Jun Hu) wrote the manuscript, and all authors read and approved it. All authors have read and agreed to the published version of the manuscript.

Funding: This research was funded by the National Natural Science Foundation of China (31871592), the Creative Research Groups of the Natural Science Foundation of Hubei Province (2020CFA009), and the Fundamental Research Funds for the Central Universities (2042022kf0015).

Institutional Review Board Statement: Not applicable.

Informed Consent Statement: Not applicable.

Data Availability Statement: The sequences of genes and proteins of rice and *Arabidopsis* mentioned in our study are available for download from the public database mentioned above.

Conflicts of Interest: The authors declare no conflict of interest.

References

1. Peng, J.; Harberd, N.P. The role of GA-mediated signalling in the control of seed germination. *Curr. Opin. Plant Biol.* **2002**, *5*, 376–381. [CrossRef]
2. Zhang, N.; Xie, Y.D.; Guo, H.J.; Zhao, L.S.; Xiong, H.C.; Gu, J.Y.; Li, J.H.; Kong, F.Q.; Sui, L.; Zhao, Z.W.; et al. Gibberellins regulate the stem elongation rate without affecting the mature plant height of a quick development mutant of winter wheat (*Triticum aestivum* L.). *Plant Physiol. Biochem.* **2016**, *1072*, 28–236. [CrossRef] [PubMed]
3. Lo, S.F.; Yang, S.Y.; Chen, K.T.; Hsing, Y.I.; Zeevaart, J.A.; Chen, L.J.; Yu, S.M. A novel class of gibberellin 2-oxidases control semidwarfism, tillering, and root development in rice. *Plant Cell* **2008**, *20*, 2603–2618. [CrossRef] [PubMed]
4. Sakata, T.; Oda, S.; Tsunaga, Y.; Shomura, H.; Kawagishi-Kobayashi, M.; Aya, K.; Saeki, K.; Endo, T.; Nagano, K.; Kojima, M.; et al. Reduction of gibberellin by low temperature disrupts pollen development in rice. *Plant Physiol.* **2014**, *164*, 2011–2019. [CrossRef]
5. Fuchs, E.; Atsmon, D.; Halevy, A.H. Adventitious staminate flower formation in gibberellin treated gynoecious cucumber plants. *Plant Cell Physiol.* **1977**, *18*, 1193–1201. [CrossRef]
6. MacMillan, J. Occurrence of Gibberellins in Vascular Plants, Fungi, and Bacteria. *J. Plant Growth Regul.* **2001**, *20*, 387–442. [CrossRef]
7. Hedden, P.; Thomas, S.G. Gibberellin biosynthesis and its regulation. *Biochem. J.* **2012**, *444*, 11–25. [CrossRef]
8. Olszewski, N.; Sun, T.P.; Gubler, F. Gibberellin signaling: Biosynthesis, catabolism, and response pathways. *Plant Cell* **2002**, *14*, S61–S80. [CrossRef]
9. Hedden, P. Gibberellin Metabolism and Its Regulation. *J. Plant Growth Regul.* **2001**, *20*, 317–318. [CrossRef]
10. Arabidopsis Genome, I. Analysis of the genome sequence of the flowering plant Arabidopsis thaliana. *Nature* **2000**, *408*, 796–815. [CrossRef]
11. Hedden, P.; Kamiya, Y. GIBBERELLIN BIOSYNTHESIS: Enzymes, Genes and Their Regulation. *Annu. Rev. Plant Physiol. Plant Mol. Biol.* **1997**, *484*, 31–460. [CrossRef] [PubMed]
12. De Carolis, E.; De Luca, V. 2-oxoglutarate-dependent dioxygenase and related enzymes: Biochemical characterization. *Phytochemistry* **1994**, *36*, 1093–1107. [CrossRef]
13. Monna, L.; Kitazawa, N.; Yoshino, R.; Suzuki, J.; Masuda, H.; Maehara, Y.; Tanji, M.; Sato, M.; Nasu, S.; Minobe, Y. Positional cloning of rice semidwarfing gene, sd-1: Rice "green revolution gene" encodes a mutant enzyme involved in gibberellin synthesis. *Dna Res. Int. J. Rapid Publ. Rep. Genes Genomes* **2002**, *9*, 11–17. [CrossRef]
14. Khush, G.S. Origin, dispersal, cultivation and variation of rice. *Plant Mol. Biol.* **1997**, *35*, 25–34. [CrossRef] [PubMed]

15. Huang, J.; Tang, D.; Shen, Y.; Qin, B.; Hong, L.; You, A.; Li, M.; Wang, X.; Yu, H.; Gu, M.; et al. Activation of gibberellin 2-oxidase 6 decreases active gibberellin levels and creates a dominant semi-dwarf phenotype in rice (*Oryza sativa* L.). *J. Genet. Genom.* **2010**, *37*, 23–36. [CrossRef]

16. Oikawa, T.; Koshioka, M.; Kojima, K.; Yoshida, H.; Kawata, M. A role of OsGA20ox1, encoding an isoform of gibberellin 20-oxidase, for regulation of plant stature in rice. *Plant Mol. Biol.* **2004**, *55*, 687–700. [CrossRef]

17. Itoh, H.; Ueguchi-Tanaka, M.; Sentoku, N.; Kitano, H.; Matsuoka, M.; Kobayashi, M. Cloning and functional analysis of two gibberellin 3 beta -hydroxylase genes that are differently expressed during the growth of rice. *Proc. Natl. Acad. Sci. USA* **2001**, *98*, 8909–8914. [CrossRef]

18. Goodstein, D.M.; Shu, S.; Howson, R.; Neupane, R.; Hayes, R.D.; Fazo, J.; Mitros, T.; Dirks, W.; Hellsten, U.; Putnam, N.; et al. Phytozome: A comparative platform for green plant genomics. *Nucl. Acids Res.* **2012**, *40*, D1178–D1186. [CrossRef]

19. Letunic, I.; Bork, P. Interactive Tree of Life (iTOL) v4: Recent updates and new developments. *Nucl. Acids Res.* **2019**, *47*, W256–W259. [CrossRef]

20. Vilella, A.J.; Severin, J.; Ureta-Vidal, A.; Heng, L.; Durbin, R.; Birney, E. Ensembl Compara Gene Trees: Complete, duplication-aware phylogenetic trees in vertebrates. *Genome Res.* **2009**, *19*, 327–335. [CrossRef]

21. Guo, A.Y.; Zhu, Q.H.; Chen, X.; Luo, J.C. GSDS: A gene structure display server. *Yi Chuan = Hered.* **2007**, *29*, 1023–1026. [CrossRef]

22. Bailey, T.L.; Boden, M.; Buske, F.A.; Frith, M.; Grant, C.E.; Clementi, L.; Ren, J.; Li, W.W. Noble WS: MEME SUITE: Tools for motif discovery and searching. *Nucl. Acids Res.* **2009**, *37*, W202–W208. [CrossRef] [PubMed]

23. Deng, W.; Wang, Y.; Liu, Z.; Cheng, H.; Xue, Y. HemI: A toolkit for illustrating heatmaps. *PLoS ONE* **2014**, *9*, e111988. [CrossRef]

24. Livak, K.J.; Schmittgen, T.D. Analysis of relative gene expression data using real-time quantitative PCR and the 2(-Delta Delta C(T)) Method. *Methods* **2001**, *25*, 402–408. [CrossRef] [PubMed]

25. Han, F.; Zhu, B. Evolutionary analysis of three gibberellin oxidase genes in rice, Arabidopsis, and soybean. *Gene* **2011**, *473*, 23–35. [CrossRef]

26. Horton, P.; Park, K.J.; Obayashi, T.; Fujita, N.; Harada, H.; Adams-Collier, C.J.; Nakai, K. WoLF PSORT: Protein localization predictor. *Nucl. Acids Res.* **2007**, *35*, W585–W587. [CrossRef] [PubMed]

27. Emanuelsson, O.; Brunak, S.; von Heijne, G.; Nielsen, H. Locating proteins in the cell using TargetP, SignalP and related tools. *Nat. Protoc.* **2007**, *2*, 953–971. [CrossRef] [PubMed]

28. Shan, C.; Mei, Z.; Duan, J.; Chen, H.; Feng, H.; Cai, W. OsGA2ox5, a gibberellin metabolism enzyme, is involved in plant growth, the root gravity response and salt stress. *PLoS ONE* **2014**, *9*, e87110. [CrossRef]

29. Long, M.; Rosenberg, C.; Gilbert, W. Intron phase correlations and the evolution of the intron/exon structure of genes. *Proc. Natl. Acad. Sci. USA* **1995**, *92*, 12495–12499. [CrossRef]

30. Kolkman, J.A.; Stemmer, W.P. Directed evolution of proteins by exon shuffling. *Nat. Biotechnol.* **2001**, *19*, 423–428. [CrossRef]

31. Morgante, M.; Brunner, S.; Pea, G.; Fengler, K.; Zuccolo, A.; Rafalski, A. Gene duplication and exon shuffling by helitron-like transposons generate intraspecies diversity in maize. *Nat. Genet.* **2005**, *37*, 997–1002. [CrossRef] [PubMed]

32. Martinez, M.; Abraham, Z.; Carbonero, P.; Diaz, I. Comparative phylogenetic analysis of cystatin gene families from arabidopsis, rice and barley. *Mol. Genet. Genom. Mgg* **2005**, *273*, 423–432. [CrossRef] [PubMed]

33. Pimenta Lange, M.J.; Liebrandt, A.; Arnold, L.; Chmielewska, S.M.; Felsberger, A.; Freier, E.; Heuer, M.; Zur, D.; Lange, T. Functional characterization of gibberellin oxidases from cucumber, *Cucumis sativus* L. *Phytochemistry* **2013**, *906*, 2–69. [CrossRef] [PubMed]

34. Lange, T. Cloning gibberellin dioxygenase genes from pumpkin endosperm by heterologous expression of enzyme activities in *Escherichia coli*. *Proc. Natl. Acad. Sci. USA* **1997**, *94*, 6553–6558. [CrossRef]

35. Salazar-Cerezo, S.; Martinez-Montiel, N.; Garcia-Sanchez, J.; Perez, Y.T.R.; Martinez-Contreras, R.D. Gibberellin biosynthesis and metabolism: A convergent route for plants, fungi and bacteria. *Microbiol. Res.* **2018**, *2088*, 5–98. [CrossRef]

36. Zhao, Z.; Zhang, Y.; Liu, X.; Zhang, X.; Liu, S.; Yu, X.; Ren, Y.; Zheng, X.; Zhou, K.; Jiang, L.; et al. A role for a dioxygenase in auxin metabolism and reproductive development in rice. *Dev. Cell* **2013**, *27*, 113–122. [CrossRef]

37. Hedden, P.; Phillips, A.L. Gibberellin metabolism: New insights revealed by the genes. *Trends Plant Sci.* **2000**, *5*, 523–530. [CrossRef]

38. Sakamoto, T.; Morinaka, Y.; Ishiyama, K.; Kobayashi, M.; Itoh, H.; Kayano, T.; Iwahori, S.; Matsuoka, M.; Tanaka, H. Genetic manipulation of gibberellin metabolism in transgenic rice. *Nat. Biotechnol.* **2003**, *21*, 909–913. [CrossRef]

Article

QTL Analysis and Heterosis Loci of Effective Tiller Using Three Genetic Populations Derived from *Indica-Japonica* Crosses in Rice

Xiaoxiao Deng [1,†], Jingzhang Wang [1,2,†], Xuhui Liu [1], Jian Yang [1], Mingao Zhou [1], Weilong Kong [1,3], Yifei Jiang [1], Shiming Ke [1], Tong Sun [1] and Yangsheng Li [1,*]

1   State Key Laboratory of Hybrid Rice, Key Laboratory for Research and Utilization of Heterosis in Indica Rice, Ministry of Agriculture, College of Life Sciences, Wuhan University, Wuhan 430072, China
2   Wuhan Wuda-Tianyuan Biotechnology Co., Ltd., Wuhan 430072, China
3   Shenzhen Branch, Guangdong Laboratory for Lingnan Modern Agriculture, Genome Analysis Laboratory of the Ministry of Agriculture, Agricultural Genomics Institute at Shenzhen, Chinese Academy of Agricultural Sciences, Shenzhen 518120, China
*   Correspondence: lysh2001@whu.edu.cn; Tel.: +86-186-0710-3953
†   These authors contributed equally to this work.

Abstract: Effective panicle numbers (PNs) and Tiller numbers (TNs) are important traits affecting rice (*Oryza sativa* L.) architecture and grain yield. However, the molecular mechanisms underlying PN and TN heterosis remain unknown in rice. In addition, new PN- or TN-related genes need to be detected and discovered. In this study, in order to detect rice quantitative trait loci (QTLs) and the heterosis-related loci of PN or TN in rice, we developed a high generation recombinant inbred line (RIL) population from a cross of two elite cultivars, Luohui9 (*Xian/Indica*) and RPY geng (*Geng/Japonica*), and two testcross hybrid populations derived from the crosses of RILs and two cytoplasmic male sterile lines, YTA (*Xian/Indica*) and Z7A (*Geng/Japonica*). Finally, nine QTLs of PN across four seasons were identified, and two QTLs of TN in 191HB were mapped. Besides this, six heterosis-related QTLs of PN and five heterosis-related QTLs of TN were located. We found that heterosis-related QTLs of PN or TN covered multiple known genes, such as *MOC1*, *TAC1* and *OsETR2*. Furthermore, homologous gene analysis identified one candidate gene of PN (*LOC_10g25720*). Together, these findings uncover multiple heterosis-related loci, and provide a new insight into the heterosis mechanism of PN and TN in rice.

Keywords: rice; effective panicle numbers; tiller numbers; QTL analysis; heterosis-related loci

Citation: Deng, X.; Wang, J.; Liu, X.; Yang, J.; Zhou, M.; Kong, W.; Jiang, Y.; Ke, S.; Sun, T.; Li, Y. QTL Analysis and Heterosis Loci of Effective Tiller Using Three Genetic Populations Derived from *Indica-Japonica* Crosses in Rice. *Agronomy* **2022**, *12*, 2171. https://doi.org/10.3390/agronomy12092171

Academic Editor: Manish K. Pandey

Received: 18 August 2022
Accepted: 7 September 2022
Published: 13 September 2022

Publisher's Note: MDPI stays neutral with regard to jurisdictional claims in published maps and institutional affiliations.

1. Introduction

Rice (*Oryza sativa* L.) is a stable food and one of the major crops in the world. Owing to the impacts of population growth and limited arable land, breeders and scientists face the challenge of cultivating higher-yield potential crops. Rice yield is a complex agronomic trait composed of four main factors, including PN, grain number per panicle, seed setting rate and 1000 grain weight. PN is an important component of grain yield. Stable PN is one of the most important characteristics of ideal plant architecture [1]. Exploring the molecular genetic mechanisms of PN is a key strategy to increase grain yield.

The dynamic change in tiller numbers (TNs) may determine the final PN [2], and PN is highly associated with TN. TN is one of most unstable and complex agronomic traits controlled by multiple genes. In the past few decades, the genetic dissection of TN by high-density molecular marker linkage mapping has uncovered a large number of QTLs. The meta-analysis of QTLs was used to merge multiple QTLs from different rice genetic populations and identify consensus and stable QTLs, which enhanced the reliability of, and narrowed down, the confidence interval [3]. Given its high reliability, Meta-QTL (MQTL) has been widely used in crop breeding [4–8]. In total, 77 QTLs for TN, published from 1996 to 2019, been analyzed, resulting in 10 MQTLs in rice [9].

A tiller is mainly formed via two distinct processes: the formation of an axillary bud, and its subsequent outgrowth [10]. Several key genes controlling tiller and panicle numbers have been cloned through the isolation of rice tiller formation mutants, such as *MOC1*, *MOC2*, *LAX1* and *LAX2* [11–13]. As the first key gene controlling rice tillering, *MOC1* encodes a GRAS family nuclear protein, and is mainly expressed in axillary buds, moreover, *MOC1* initiates axillary bud growth and promotes their outgrowth in rice [14]. In addition, *MOC1* acts as a co-activator of *MOC3*, and *MOC3* can directly bind the promoter of FLORAL ORGAN NUMBER1 (*FON1*). *MOC1* and *MOC3* physically interact to regulate tiller bud outgrowth via the expression of *FON1* [12]. *MOC2* encodes cytosolic fructose-1,6-bisphosphatase 1 (FBP1) and promotes tiller bud outgrowth through participating in the sucrose biosynthesis pathway [15]. *OsNAC23*, a sugar-inducible NAC transcription factor, directly binds the promoter of the Tre6P phosphatase gene *TPP1* and elevates Tre6P in rice. Furthermore, overexpressing *OsNAC23* in three different rice cultivar backgrounds increases the Tre6P content, and enables the accumulation of more panicle numbers and a higher yield [16]. Moreover, few TN- or PN-related genes have been cloned through the mapping of genetic populations [17].

As a major determinant of rice architecture, rice tillering is found to be associated with multiple plant hormones, such as brassinosteroids (BRs), auxins, cytokinins (CKs), strigolactones (SLs), and ethylene. *D10* encodes carotenoid cleavage dioxygenase 8 and affects rice tillering through participating in strigolactone biosynthesis; the expression of *D10* is induced by the exogenous auxin [18], which indicates that rice tillering is affected by both auxin and SLs. Moreover, SLs integrate with CKs to regulate rice tillering. The interaction among SLs, CKs, and auxin controls rice tillering [19].

In this study, Luohui9 was found to be an elite *Xian/Indica* cultivar, and RPY geng was an elite *Geng/Japonica* cultivar. The F_1 progeny of Luohui9 and RPY gneg exhibit obvious heterosis in terms of PN. In order to detect QTLs for PN and TN, and uncover the heterosis loci for PN and TN, a high-generation RIL population derived from the cross between Luohui9 and RPY geng was developed, and two testcross hybrid populations derived from the crosses of RILs and two cytoplasmic male sterile lines, YTA (*Xian/Indica*) and Z7A (*Geng/Japonica*), were constructed. A high-density bin map was constructed [20], and the genomes of Luohui9 and RPY geng were de novo assembled [6]. Nine QTLs of PN were detected in RILs, and six heterosis-related QTLs of PN were mapped in two testcross populations. Besides this, homologous blast identified new candidate genes in a novel QTL cluster. This study used three genetic populations to identify multiple QTLs and heterosis-related QTLs for PN, and provided new insight into the heterosis mechanism of PN in rice.

2. Materials and Methods

2.1. Plant Materials and Population Construction

Luohui9 has been used as the female parent to cross with RPY geng since 2011; the F_1 plant was self-crossed, and then inbred over 10 generations by single-seed breeding to generate an RIL population containing 272 inbred lines [20,21]. To detect PN and TN heterosis-related QTLs, we used the RILs (F_{14}) to cross with an *Xian/Indica*-type cytoplasmic male sterile line, YTA and a *Geng/Japonica*-type cytoplasmic male sterile line, ZTA, respectively, and generated two testcross hybrid populations, 209 YTA-TCF$_1$ and 173 Z7A-TCF$_1$.

In 2017–2018, RILs (F_{11}, F_{12}, F_{14}, and F_{16}) and their parents were planted in the experimental fields of the Ezhou Rice Breeding Experimental Base of Wuhan University in summer and autumn, and at the Rice Experimental Base of Tianyuan Co., Ltd. in Lingshui County, Hainan Province, in winter and spring every year. F_{11} was placed in HN (abbreviated as HN162), F_{12} in HB (abbreviated as HB171), F_{14} in HB (abbreviated as HB181), and F_{16} in HB (abbreviated as HB191). Two testcross hybrid populations and their parents were planted at the Hannan Breeding Experimental Base of Tianyuan Co., Ltd. in Hannan District, Wuhan City, Hubei Province from May to October 2019. All plants were planted under standard agricultural planting management procedures. A randomized complete block design was employed for RILs, and two testcross populations and 15 replications for Luohui9 and RPY geng were used to evaluate the overall uniformity of the experimental field.

2.2. Trait Statistics

Sixty plants of each line from RILs and two testcross populations were planted in six rows, with ten plants per row. The spacing of plants and lines was 13.3 cm × 20 cm and 30 cm, respectively. The PN of RILs was surveyed in the four environments (HB191, HB181, HB171, and HN162). The TN of RILs was surveyed in the HB191 environment. The PN and TN of two testcross populations (YTA-TCF$_1$ and Z7A-TCF$_1$ population) were investigated in the HB191 environment. Each inbred line was used to plant 60 individual plants. For PN, 5 individual plants were chosen and counted, and the average value of 5 individual plants of PN was considered as the value of the inbred line of PN. For TN, 9 individual plants were chosen and counted, and the average value of 9 individual plants of TN was considered as the value of the inbred line of TN.

2.3. Bin Mapping-Based QTL Analysis

The genetic linkage maps of 272 RILs, including 4578 bin blocks with the total bin-map distance of 2356.41 cm, were previously constructed in our lab [21]. The QTL mapping of PN and TN was performed in the R package "R/qtl" [22], the CIM interval mapping method was adopted, and the LOD threshold was set at 2.5. The confidence interval was calculated with the function "lodint" [23]. The drop value was set to 1.5. We filtered QTLs with regions greater than 7 Mb. The MQTLs of TN were derived from [6]. The visualization distributions of QTLs and PN-related known genes in the chromosome were generated by the Mapchart software [24].

2.4. Heterosis Analysis

For the mapping of heterosis-related loci, the MH, TII, and PH values were calculated by the formulas: MH = [F1 − (P1 + P2)/2]/[(P1 + P2)/2]; LH = [F1 − P2]/P2(P1 > P2); TH = [F1 − P1]/P1(P1 > P2) [25]. MH represents middle-parent heterosis, LH represents lower-parent heterosis, and TH represents transgressive heterosis, while P1 and P2 represent the parents. The QTL mapping of PN and TN heterosis-related indexes was performed using the R package "R/qtl" [22], while PN and TN heterosis-related indexes were counted by the MH, TH, and PH values with the formula $y = 2^x$. The CIM interval mapping method was adopted and the LOD threshold was set at 2.5. The confidence interval was calculated with the function "lodint" [23]. The drop value was set at 1.5. We filtered QTLs with regions greater than 7 Mb.

2.5. Candidate Gene Prediction

Because most homologous genes have the same or similar functions, the protein sequences of 157 PN-related known genes were collected from http://rice.plantbiology.msu.edu/pub/data/Eukaryotic_Projects/o_sativa/annotation_bs/pseudomolecules/version_7.0/all.dir/ (accessed on 8 October 2021) as the query sequences for blast analysis. We extracted the protein sequence of the genes contained in the QTL region as blastdb. Homologous genes were identified with blast P with an evalue 1×10^{-20}. The protein sequence of the homologous genes was extracted from the RPY geng and Luohui9 genome documents for further sequence alignments by DNAman. Homologous genes with no difference between the parents were not considered as candidate genes. Further, the PNs for RILs with male parental genotypes and with female parent genotypes in multiple environments were compared. The homologous genes for which the male parental genotype and the female parental genotype manifested significant differences in terms of PNs for RILs in all four different environments were consider as the candidate genes [7,26].

3. Results

3.1. The Trait Performance of Effective Panicle Numbers per Plant (PN) from the RILs and the Testcross Hybrid Populations

Here, the PNs of RILs were investigated in Hubei (HB) or Hainan (HN) over four growing seasons. The PNs of the testcross hybrid populations were investigated in Hubei in 2019. The PNs in RILs and two testcross populations all showed normal distributions, indicating that PN and PN heterosis-related traits were typical quantitative traits involving multiple genes (Figure 1). The PN of Luohui9 was more than that of RPY geng. The phenotypic variation and average of the PN trait from RILs and two testcross populations were compared in HB191. The PN of RILs mainly ranged from 3.20 to 11.80, while that of the YTA-TCF$_1$ population ranged from 4.00 to 20.00, and that of the ZTA-TCF$_1$ population ranged from 7.00 to 20.00. The variations in PN in the testcross hybrid populations were larger than those in the RILs. Besides this, the average values of PN for RILs were significantly lower than those of the YTA-TCF$_1$ population and the Z7A-TCF$_1$ population (Supplemental Table S1).

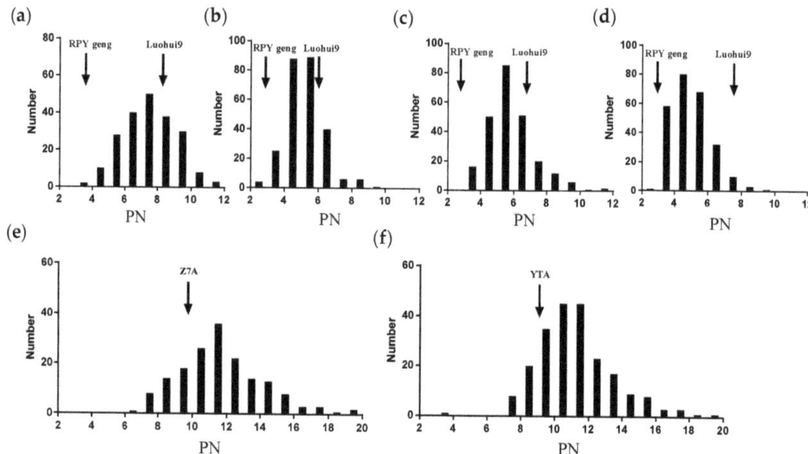

Figure 1. The distribution pattern of the PN trait in RILs and two testcross hybrid populations. (**a–d**) Column graphs of PN for RIL population in HB191 (**a**), HB181 (**b**), HB171 (**c**), and HN162 (**d**), respectively. (**e,f**) Column graph of PN for Z7A-TCF$_1$ population (**e**) and YTA-TCF$_1$ population (**f**) in HB191.

3.2. QTLs Detection in RILs and Heterosis-Related QTLs Detection in Two Testcross Hybrid Populations

Nine QTLs of PN were identified, explaining 4.83–10.80% of the phenotypic variation. Importantly, two QTLs, namely, qPN-10-2 and qPN-10-3, had a common region, with a size of 108 kb. In addition, qPN-10-2 and qPN-10-3 were detected in HB171 and HN162 separately, explaining 6.31% and 9.49% of the phenotypic variation (Supplemental Table S2).

Two testcross populations were used to detect heterosis-related QTLs of PN. A total of six heterosis-related QTLs of PN were mapped, explaining 4.76–11.11% of the phenotypic variation. Those six heterosis-related QTLs included a transgressive heterosis-related QTL (*qTH-PN-2-1*), three mid-parent heterosis-related QTLs (*qMH-PN-2-1*, *qMH-PN-9-1* and *qMH-PN-6-1*), and two low-parent heterosis-related QTLs (*qLH-PN-4-1* and *qLH-PN-2-1*) (Supplemental Table S3). No heterosis-related QTLs overlapped with QTLs for PN, which indicated that both heterosis-related QTLs and QTL for PN were independent.

3.3. Known Genes Affecting the PN of RILs in QTLs Interval

In this present study, 157 known PN-related genes were retrieved and compared with the position of the QTLs detected in this study. Two known genes, namely, *OsNAC23* and

OsCDC48, were covered by the QTLs of PN (Figure 2). *OsCDC48* was located in *qPN-3-2*, and the major QTL explained 10.80% of the phenotypic variation. In addition, *OsNAC23* was located in *qPN-2-1*, which explained 8.80% of the phenotypic variation. We compared PN with male/female parental genotypes of *OsNAC23*. The male parental genotype and female parental genotype of *OsNAC23* for PN in RILs showed significant differences in the HB191 and HN162 environments (Figure 3a), suggesting that *OsNAC23* had a great influence on PN. Besides this, the comparison of the PN of RILs with parental genotypes of *OsCDC48* showed significant differences in the HB171 and HN162 environments (Figure 3b), indicating that *OsCDC48* had a great influence on PN.

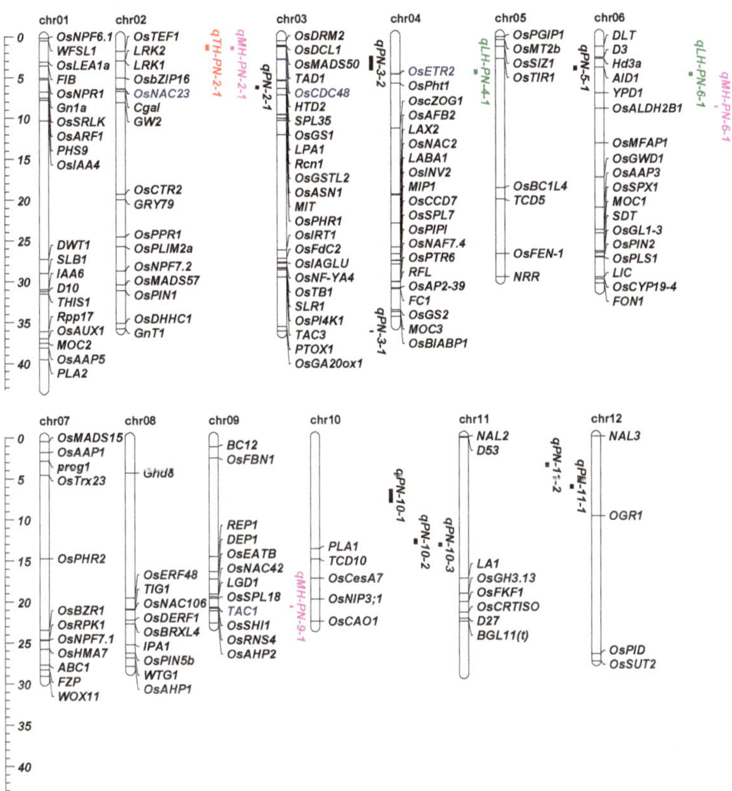

Figure 2. The position of quantitative trait locus (QTLs) for effective panicle numbers (PN) in recombinant inbred lines (RILs) and heterosis-related QTLs in two testcross hybrid populations, and the location comparison of them and PN-related known genes. The QTLs and heterosis-related QTLs located in this study are labeled on the right side of the chromosome; black represents the QTLs. Green, pink and red represent the relative low-parent, middle-parent and transgressive heterosis-related QTLs, respectively. The PN-related known genes were located in the chromosome. Blue represents the known genes covered by QTLs of PN, or heterosis. The Y-axis represents the physical distance (Mb) of the chromosomes.

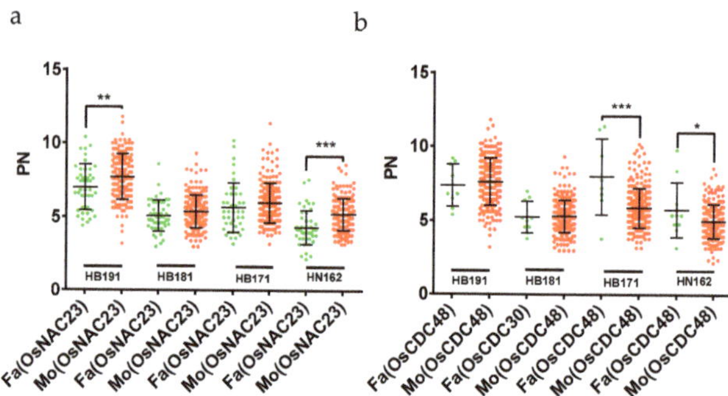

Figure 3. PN trait comparison of RILs with parental genotypes of two known genes in multiple environments. (**a**) PN trait comparison of RILs with parental genotypes of *OsNAC23* in multiple environments (HB191, HB181, HB171, HN162); (**b**) PN trait comparison of RILs with parental genotypes of *OsCDC48* in multiple environments (HB191, HB181, HB171, HN162). RILs with the male parental genotype and female parental genotype of the *OsNAC23* and *OsCDC48* are marked by green dots and red dots in point clouds, respectively. * represents $p < 0.05$, ** represents $p < 0.01$, *** represents $p < 0.001$. PN: the effective panicle numbers per plant.

3.4. Kown Genes Acted as Heterosis-Related QTLs for PN

Six heterosis-related QTLs for PN were detected in two testcross populations, four heterosis-related QTLs were detected in the YTA-TCF$_1$ population, and two heterosis-related QTLs were detected in the Z7A-TCF$_1$ population. Among the 157 known PN-related genes, two genes (*OsETR2* and *TAC1*) overlapped with heterosis-related QTL for PN. *OsETR2* was located in *qLH-PN-4-1*, which was mapped in the YTA-TCF$_1$ population. *OsETR2* was mapped on chromosome 4 and acted as the ethylene receptor, encoding serine/threonine kinase. The effective panicles were reduced in the OsETR2-overexpressing plants [27]. Interestingly, in our previous research, we detected a grain-shaped heterosis-related QTL *qTH-GLWR-4-1* in rice using the same YTA-TCF$_1$ population, and found that *qTH-GLWR-4-1* covered *OsETR2* [21], which suggests that *OsETR2* may play a vital role in the heterosis of yield-related traits, such as PN and grain shape. In addition, *TAC1* was located in *qMH-PN-9-1*, which was mapped in the YTA-TCF$_1$ population. *TAC1* was a major QTL controlling the tiller angle in rice [28]. In this study, *TAC1* was found in the PN heterosis-related QTL, and may result in the heterosis of PN. Whether *OsETR2* and *TAC1* play a vital role in heterosis remains to further explored.

3.5. Prediction of Candidate Gene in the QTL Cluster

We focused on the QTL cluster in chromosome 10, including *qPN-10-2*, and *qPN-10-3*. *qPN-10-2* and *qPN-10-3* were mapped in HB171 and HN162 separately. Besides this, *qPN-10-2* and *qPN-10-3* had a common QTL region. Interestingly, our other study also detected a QTL for other tiller numbers, namely, *qTN-10-2* (10029667–13469285, 3.44 Mb), which completely covered *qPN-10-2*, in HN162 [20]. Among these, the overlap interval (13049716–13469285, only 0.42 Mb) of two QTLs (*qPN-10-2*, and *qTN-10-2*) was just 0.42 Mb, which indicates that the stable QTL in chromosome 10 controls TN and PN in both the HB environment and the HN environment. Moreover, no known PN-related genes were located in the QTL cluster. The common regions of *qPN-10-2* and *qTN-10-2* harbored 70 candidate genes. To search for new PN-related genes in the QTL cluster, 157 known PN-related genes were selected to search for homologous genes. Based on homolog identification, three genes (*LOC_10g25720*, *LOC_Os10g25830*, and *LOC_Os10g25890*) were obtained. The comparison of the differences between the male parental protein sequence and the female parental protein sequence of these three genes showed that *LOC_10g25720* and *LOC_Os10g25830* had such differences, while *LOC_Os10g25890* had no differences between male parental protein sequence and female parental sequence (Table 1). The PNs for RILs with the male parental genotype and the female parent genotype of these three genes were compared; only *LOC_10g25720* showed significant differences in all four different environments, while *LOC_Os10g25830* and *LOC_Os10g25890* only showed significant differences in HB171, and there were no significant findings in the other three environments (Figure 4). So, *LOC_10g25720* was predicted to be the only candidate gene in the QTL cluster in chromosome 10. In addition, *LOC_10g25720* was the *D10* homologous gene.

Table 1. Alignment results of parental protein sequences of three genes in *qPN-10-2*.

Gene ID	Gene Position (bp)	Homologous Known Gene	QTLs	QTL Position (bp)	Male Parental Gene/Female Parental Gene	Difference of Protein Sequence
LOC_Os10g25720	13318548-13322321	*D10*	*qPN-10-2*	13238715-13251104	FaEVM0003415.1/MoEVM0001835.1	Yes
LOC_Os10g25830	13387338-13388803	*MIT*	*qPN-10-2*	13238715-13251104	FaEVM0043545.1/MoEVM0016401.1	Yes
LOC_Os10g25890	13416622-13421264	*D10*	*qPN-10-2*	13238715-13251104	FaEVM0032270.1/MoEVM0005566.1	No

Note: Yes means there is a difference in the parent protein sequence; No means there is no difference in the parent protein sequence.

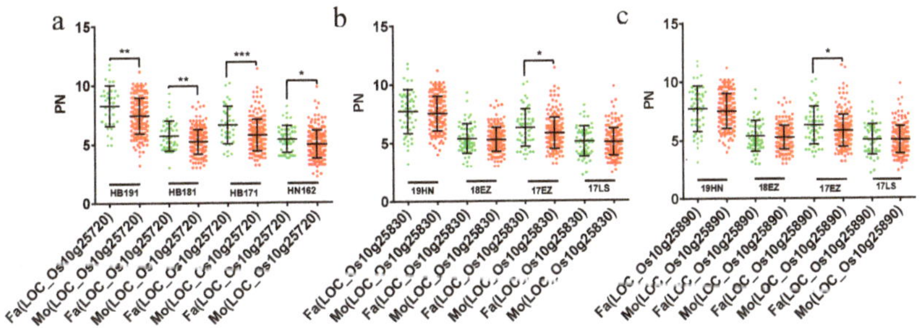

Figure 4. PN trait comparison of RILs with parental genotypes of *LOC_10g25720* (**a**), *LOC_10g25830* (**b**), and *LOC_10g25890* (**c**) in multiple environments. RILs with the male parental genotype and female parental genotype of *LOC_10g25720*, *LOC_10g25830*, and *LOC_10g25890* are marked by green dots and red dots in point clouds, respectively. * represents $p < 0.05$, ** represents $p < 0.01$, *** represents $p < 0.001$. PN: Effective panicle number.

3.6. MQTLs and QTLs of Effective Tiller and Their Heterosis-Related Loci Cover Multiple Known Genes

In the next step of analysis, we detected QTLs of TN in the 191HB environment using the RIL population. qTN-6-1 and qTN-11-1 were detected, explaining 9.41% and 5.64% of the phenotypic variation (Table S4). Besides this, the heterosis-related QTLs of TN were mapped in the 191HB environment using two testcross populations. Three heterosis-related QTLs of TN were detected in the YTA-TCF₁ population, including *qMH-TN-2-1*, *qMH-TN-6-1* and *qMH-TN-9-1*. Interestingly, *qMH-TN-6-1* covered *MOC1*. We compared the differences between the male and female parental protein sequence of *MOC1*, and found that there was no difference. Then, we further compared the difference between the male and female parental nucleotide sequences of *MOC1*, and the results show that, compared with the female parental nucleotide sequence of *MOC1*, the male parental nucleotide sequence of *MOC1* had a 10 bp deletion in the intron region (Figure S1). Whether *MOC1* plays a role in heterosis remains to be further explored. In addition, two heterosis-related QTLs of TN were detected in the Z7A-TCF₁ population, including *qTH-TN-2-1* and *qLH-TN-2-1* (Table S5). Next, we collected the QTLs, the heterosis-related QTLs of PN and TN, QTLs of TN [20], and MQTLs of TN [9], and compared their loci (Figure 5). *qTN-2-1* covered *OsPPR1* and *OsPLIM2a*. Overexpressing *OsPLIM2a* can result in reducing rice tillers [29]. MQTL-TN2 covered *OsIAA6* and *D10*. Besides this, MQTL-TN9 covered *OsERF48* and *TIG1*. MQTL-TN10 covered *LGD1* and *OsSPL18*. Moreover, *LPA1* and *OsPIPI* were located in MQTL-TN4 and MQTL-TN4 separately. *OsNF-YB9* was covered by *qTN-6-1*, and overexpressing *OsNF-YB9* can result in increasing numbers of effective panicles [30].

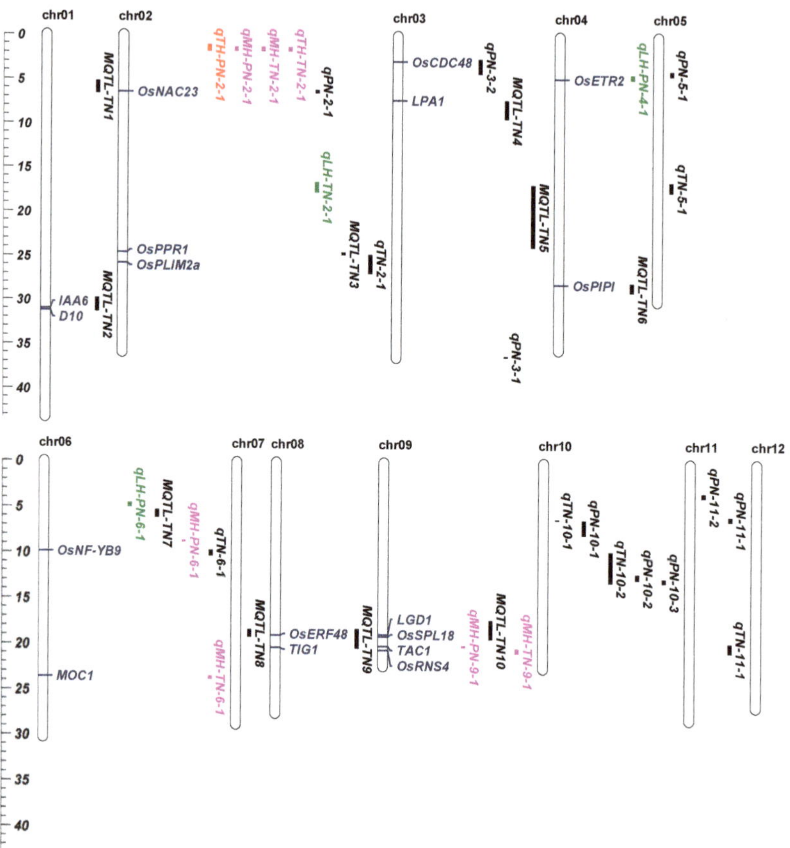

Figure 5. The position of the quantitative trait locus (QTLs) for effective panicle numbers (PN), tiller numbers (TN) in recombinant inbred lines (RILs) and heterosis-related QTLs in two testcross hybrid populations, and the location comparison of them and MQTLs. The QTLs and heterosis-related QTLs located in this study are labeled on the right side of the chromosome; black represents the QTLs and MQTLs. Green, pink, and red represent the relative low-parent, middle-parent and transgressive heterosis-related QTLs, respectively. Blue represents the known genes, covered by QTLs, MQTLs and heterosis-related QTLs. The Y-axis represents the physical distance (Mb) of the chromosomes.

4. Discussion

TN determines PN, and largely affects grain yield. Therefore, TN is one of the most important agronomic traits controlled by multiple genes. TN- or PN-related genes were detected through the mapping of RIL populations. It is reported that *OsNAC23* affects rice grain size and grain weight [31]. A recent study reveal that Tre6P, *OsNAC23* and SnRK1a physically interact to regulate sugar homeostasis and grain yield in rice [16]. In addition, Tre6P is a central sugar signal in plants, and *OsNAC23* directly binds the promoter of the Tre6P phosphatase gene *TPP1*, while Tre6P has great potential to improve crop yield. The heterologous expression of the *OsTPP1* in maize directly increased the yield by 9–49% [32]. The spraying of absorbable Tre6P precursor also increases wheat yield by 20% [33]. These studies indicate that *OsNAC23* acts as a sugar indicator, and senses the availability of sugar for promoting rice growth. Moreover, *OsNAC23* affected multiple rice agronomic traits, including grain size, grain weight and grain yield. However, whether *OsNAC23* affected TN or PN is still unknown. We have observed that *qPN-2-1* was detected using the RIL population, which contained *OsNAC23*. The male parental genotype and female parental genotype of *OsNAC23* for PN in RILs showed significant differences between the HB191 and HN162 environments, suggesting that *OsNAC23* had a great influence on PN. *OsNAC23* might regulate effective panicle numbers through sensing the availability of sugar and participating in sugar signaling. Besides this, MQTL-TN2 covered *D10* and *OsIAA6*. *OsIAA6* encodes the Aux/IAA protein and is involved in the control of tiller outgrowth [34]. *D10* encodes carotenoid cleavage dioxygenases and is conserved across species, such as arabidopsis, pea, and petunia, which is the rice ortholog of *MAX4/RMS1/DAD1*, and *D10*, induced by the exogenous auxin, controls rice tillering through participating in strigolactone biosynthesis and auxin signal transduction [18]. Interestingly, *D10* homologous genes *LOC_10g25720* and *LOC_Os10g25890* were covered by *qPN-10-2*, which indicates that *D10* and its homologous genes in rice may play vital roles in rice tillering through participating in strigolactone biosynthesis and auxin signaling transduction.

Discovering new PN- or TN-related genes is of great significant. PNs showed greater variation, and were more sensitive to changes in the geographical environment (Table S1). In this study, nine QTLs of PN were detected in RILs and seven QTLs (*qPN-3-1*, *qPN-5-1*, *qPN-10-1*, *qPN-10-2*, *qPN-10-3*, *qPN-11-1* and *qPN-11-2*) were newly detected. Two QTLs of PN were detected in RILs, and one QTL, *qTN-11-1*, was newly detected. *qPN-11-1* was 477.98 kb, containing 72 genes. *qPN-3-1* was just 96.19 kb, containing 15 genes. A QTL cluster was found in chromosome 10, including *qPN-10-2*, *qPN-10-3* and *qTN-10-2*. The overlap region of two QTLs (*qPN-10-2* and *qTN-10-2*) was just 0.42 Mb, containing 63 genes. Such a narrow interval allowed us to reduce the number of candidate genes through the homolog identification of known genes. Based on homolog identification, three genes (*LOC_10g25720*, *LOC_Os10g25830*, and *LOC_Os10g25890*) were obtained. Only *LOC_10g25720* showed significant differences in all of the four different environments, while *LOC_Os10g25830* and *LOC_Os10g25890* only showed significant differences in HB171, and no significant differences were found in the other three environments. PN phenotype data from multiple environments can increase the accuracy of the results used for gene genotyping.

To explore rice heterosis, lots of QTLs related to heterosis for various agronomic traits have been reported in different genetic populations [35–37], while few heterosis-related genes for PN or TN have been reported to date. In this study, we detected multiple PN and TN heterosis-related loci. An effective tiller heterosis-related QTL cluster (*qTH-PN-2-1*, *qMH-PN-2-1*, *qTH-TN-2-1*, *qMH-TN-2-1*) was found. *qTH-PN-2-1*, *qMH-PN-2-1* *qTH-TN-2-1*, and *qMH-TN-2-1* completely matched each other—the overlap region was 459.38 kb. The heterosis-related QTL *qLH-PN-4-1* contains *OsETR2*. *OsETR2* encodes serine/threonine kinase and acts as an ethylene receptor. Overexpressing *OsETR2* results in reducing the effective panicles in rice [27]. In addition, in another research, *OsETR2* and *LOC_Os04g51950* were covered by heterosis-related QTL *qTH-GLWR-4-1* and *qTH-GLWR-4-2*, separately [21]. Interestingly, *LOC_Os04g51950* also encodes serine/threonine kinase HT1. Furthermore, a recent study reported that the yields of four *TaCol-B5*-overexpressing lines of Yangmai18 were increased by an average of 11.9% compared with non-transgenic Yangmai18. Further studies have shown that *TaCol-B5* is activated by the phosphorylation of the serine/threonine protein kinase TaK4, and Ser269 of *TaCol-B5* is the phosphorylation site of *TaK4* [38]. These findings suggest that the serine/threonine protein kinase genes may play vital roles in heterosis for yield-related traits, such as grain weight, grain shape, TN and PN. Besides this, the heterosis-related QTL *qLH-PN-4-1* contained *TAC1*. *TAC1* is a major QTL controlling tiller angle [28]. No publication has reported that *TAC1* can result in heterosis for PN. In this study, *TAC1* may result in heterosis for PN, which provides a new insight into heterosis for PN in rice. *qMH-TN-6-1* covered *MOC1*. *MOC1* initiates axillary buds and affects tiller numbers in rice [14]. In this study, *MOC1* may result in heterosis for TN in rice. Interestingly, we found there was no difference between the male and female parental protein sequence of *MOC1*, while compared with the female parental nucleotide sequence of *MOC1*, the male parental nucleotide sequence had a 10 bp deletion in the intron region (Figure S1). A previous study collected 240 rice cultivars and compared the nucleotide sequences of *MOC1* for 240 rice cultivars. The results show that there was no difference in the coding region of the *MOC1* gene, while the nocoding region showed differences [39]. In addition, the *MOC1* gene was divided into six genotypes based on 21 SNP of *MOC1* promoter, and the expression level of *MOC1* with the MOC1-Hap4 genotype was slightly higher than that with the MOC1-Hap1 genotype [39]. This indicates that the *MOC1* coding sequence is extremely conserved, and the noncoding region of *MOC1* may affect the expression of *MOC1*. Therefore, we surmise that the differences in the *MOC1* noncoding region may lead to differences in the expression of *MOC1*, and ultimately affect rice tillering, resulting in heterosis for TN.

5. Conclusions

In this study, we detected nine QTLs of PN, two QTLs of TN, six heterosis-related QTLs of PN and five heterosis-related QTLs of TN. We found that heterosis-related QTLs of PN and TN covered multiple known genes, such as *MOC1*, *TAC1* and *OsETR2*. We conclude that the serine/threonine protein kinase genes may play vital roles in heterosis for yield-related traits, such as grain weight, grain shape, TN and PN. Moreover, we have predicted one PN-related candidate gene, *LOC_10g25720*. Our research uncovers multiple heterosis-related loci, and provides new insights into the heterosis mechanism of TN and PN in rice.

Supplementary Materials: The following supporting information can be downloaded at https://www.mdpi.com/article/10.3390/agronomy12092171/s1, Figure S1: Gene nucleotide sequence alignments of MOC1 in male parent, female parent and Nipponbare; Table S1: Summarized statistics for PN trait of RIL population, two testcross hybrid populations (YTA-TCF1 and Z7A-TCF1) and its parents; Table S2: QTLs for PN trait in RILs; Table S3: Low-parent, mid-parent and transgressive heterosis-related QTLs for PN trait in two testcross hybrid populations; Table S4: QTLs for TN trait in RILs; Table S5: Low-parent, mid-parent and transgressive heterosis-related QTLs for TN trait in two testcross hybrid populations.

Author Contributions: Y.L. and X.D. designed this study and developed the RIL population. X.D. and J.W. performed the experiment. X.D., W.K. and X.L. analyzed the data. X.D. wrote the manuscript. Y.L., J.Y., Y.J., M.Z., S.K. and T.S. helped in manuscript revision. X.D. and J.W. contributed equally to this work. All authors have read and agreed to the published version of the manuscript.

Funding: This study was supported by the National Key Research and Development Program of China (2016YFD0100400).

Institutional Review Board Statement: Not applicable.

Informed Consent Statement: Not applicable.

Data Availability Statement: All data generated or analyzed during this study are included in this published article (and its Supplementary Information files).

Conflicts of Interest: The authors declare no conflict of interest.

References

1. Jiao, Y.Q.; Wang, Y.H.; Xue, D.W.; Wang, J.; Yan, M.X.; Liu, G.F.; Dong, G.J.; Zeng, D.L.; Lu, Z.F.; Zhu, X.D.; et al. Regulation of OsSPL14 by OsmiR156 defines ideal plant architecture in rice. *Nat. Genet.* **2010**, *42*, 541–544. [CrossRef] [PubMed]
2. Ma, X.Q.; Li, F.M.; Zhang, Q.; Wang, X.Q.; Guo, H.F.; Xie, J.Y.; Zhu, X.Y.; Ullah Khan, N.; Zhang, Z.Y.; Li, J.J.; et al. Genetic architecture to cause dynamic change in tiller and panicle numbers revealed by genome-wide association study and transcriptome profile in rice. *Plant J.* **2020**, *104*, 1603–1616. [CrossRef] [PubMed]
3. Arcade, A.; Labourdette, A.; Falque, M.; Mangin, B.; Chardon, F.; Charcosset, A.; Joets, J. BioMercator: Integrating genetic maps and QTL towards discovery of candidate genes. *Bioinformatics* **2004**, *20*, 2324–2326. [CrossRef] [PubMed]
4. Zhang, X.C.; Shabala, S.; Koutoulis, A.; Shabala, L.; Zhou, M.X. Meta-analysis of major QTL for abiotic stress tolerance in barley and implications for barley breeding. *Planta* **2017**, *245*, 283–295. [CrossRef]

5.   Martinez, A.K.; Soriano, J.M.; Tuberosa, R.; Koumproglou, R.; Jahrmann, T.; Salvi, S.; Yield, Q.T. Lome distribution correlates with gene density in maize. *Plant Sci.* **2016**, *242*, 300–309. [CrossRef]

6.   Kong, W.L.; Deng, X.X.; Liao, Z.Y.; Wang, Y.B.; Zhou, M.A.; Wang, Z.H.; Li, Y.S. De novo assembly of two chromosome-level rice genomes and bin-based QTL mapping reveal genetic diversity of grain weight trait in rice. *Front. Plant Sci.* **2022**, *13*, 995634. [CrossRef]

7.   Kong, W.L.; Zhang, C.H.; Qiang, Y.L.; Zhong, H.; Zhao, G.Q.; Li, Y.S. Integrated RNA-seq analysis and Meta-QTLs mapping provide insight into cold stress response in rice seeding roots. *Int. J. Mol. Sci.* **2020**, *21*, 4615. [CrossRef]

8.   Kong, W.L.; Zhong, H.; Gong, Z.Y.; Fang, X.Y.; Sun, T.; Deng, X.X.; Li, Y.S. Meta-analysis of salt stress transcriptome responses in different rice genotypes at the seedling stage. *Plants* **2019**, *8*, 64. [CrossRef]

9.   Khahani, B.; Tavakol, E.; Shariati, V.; Fornara, F. Genome wide screening and comparative genome analysis for Meta-QTLs, ortho-MQTLs and candidate genes controlling yield and yield-related traits in rice. *BMC Genom.* **2020**, *21*, 294–318. [CrossRef]

10. Zhang, C.Y.; Liu, J.; Zhao, T.; Gomez, A.; Li, C.; Yu, C.S.; Li, H.Y.; Lin, J.Z.; Yang, Y.Z.; Liu, B.; et al. A drought-inducible transcription factor delays reproductive timing in rice. *Plant Physiol.* **2016**, *171*, 334–343. [CrossRef]

11. Oikawa, T.; Kyozuka, J. Two-step regulation of LAX PANICLE1 protein accumulation in axillary meristem formation in rice. *Plant Cell.* **2009**, *21*, 1095–1108. [CrossRef] [PubMed]

12. Shao, G.N.; Lu, Z.F.; Xiong, J.S.; Wang, B.; Jing, Y.H.; Meng, X.B.; Liu, G.F.; Ma, H.Y.; Liang, Y.; Chen, F.; et al. Tiller bud formation regulators MOC1 and MOC3 cooperatively promote tiller bud outgrowth by activating FON1 expression in rice. *Mol. Plant* **2019**, *12*, 1090–1102. [CrossRef] [PubMed]

13. Tabuchi, H.; Zhang, Y.; Hattori, S.; Omae, M.; Shimizu-Sato, S.; Oikawa, T.; Qian, Q.; Nishimura, M.; Kitano, H.; Xie, H.; et al. LAX PANICLE2 of rice encodes a novel nuclear protein and regulates the formation of axillary meristems. *Plant Cell.* **2011**, *23*, 3276–3287. [CrossRef] [PubMed]

14. Li, X.Y.; Qian, Q.; Fu, Z.M.; Wang, Y.H.; Xiong, G.S.; Zeng, D.; Wang, X.Q.; Liu, X.F.; Teng, S.; Hiroshi, F.; et al. Control of tillering in rice. *Nature* **2003**, *422*, 618–621. [CrossRef]

15. Koumoto, T.; Shimada, H.; Kusano, H.; She, K.C.; Iwamoto, M.; Takano, M. Rice monoculm mutation moc2, which inhibits outgrowth of the second tillers, is ascribed to lack of a fructose-1,6-bisphosphatase. *Plant Biotechnol.* **2013**, *30*, 47–56. [CrossRef]

16. Li, Z.Y.; Wei, X.J.; Tong, X.H.; Zhao, J.; Liu, X.X.; Wang, H.M.; Tang, L.Q.; Shu, Y.Z.; Li, G.H.; Wang, Y.F.; et al. The OsNAC23-Tre6P-SnRK1a feed-forward loop regulates sugar homeostasis and grain yield in rice. *Mol. Plant* **2022**, *15*, 706–722. [CrossRef]

17. Lei, L.; Zheng, H.L.; Wang, J.G.; Liu, H.L.; Sun, J.; Zhao, H.W.; Yang, L.M.; Zou, D. Genetic dissection of rice (*Oryza sativa* L.) tiller, plant height, and grain yield based on QTL mapping and metaanalysis. *Euphytica* **2018**, *214*, 109. [CrossRef]

18. Arite, T.; Iwata, H.; Ohshima, K.; Maekawa, M.; Nakajima, M.; Kojima, M.; Sakakibara, H.; Kyozuka, J. DWARF10, an RMS1/MAX4/DAD1 ortholog, controls lateral bud outgrowth in rice. *Plant J.* **2007**, *51*, 1019–1029. [CrossRef]

19. Zha, M.; Imran, M.; Wang, Y.; Xu, J.; Ding, Y.; Wang, S.H. Transcriptome analysis revealed the interaction among strigolactones, auxin, and cytokinin in controlling the shoot branching of rice. *Plant Cell Rep.* **2019**, *38*, 279–293. [CrossRef]

20. Kong, W.L.; Deng, X.X.; Yang, J.; Zhang, C.H.; Sun, T.; Ji, W.J.; Zhong, H.; Fu, X.P.; Li, Y.S. High-resolution bin-based linkage mapping uncovers the genetic architecture and heterosis-related loci of plant height in indica-japonica derived populations. *Plant J.* **2022**, *110*, 814–827. [CrossRef]

21. Deng, X.X.; Kong, W.L.; Sun, T.; Zhang, C.H.; Zhong, H.; Zhao, G.Q.; Liu, X.H.; Qiang, Y.L.; Li, Y.S. Bin mapping-based QTL analyses using three genetic populations derived from indica-japonica crosses uncover multiple grain shape heterosis-related loci in rice. *Plant Genome* **2022**, *15*, e20171. [PubMed]

22. Arends, D.; Prins, P.; Jansen, R.C.; Broman, K.W. R/qtl: High-throughput multiple QTL mapping. *Bioinformatics* **2010**, *26*, 2990–2992. [CrossRef] [PubMed]

23. Dupuis, J.; Siegmund, D. Statistical methods for mapping quantitative trait loci from a dense set of markers. *Genetics* **1999**, *151*, 373–386. [CrossRef] [PubMed]

24. Voorrips, R.E. Mapchart: Software for the graphical presenttion of linkage maps and QTLs. *J. Hered.* **2002**, *93*, 77–78.

25. Patrick, S.S.; Nathan, M.S. Progress toward understanding heterosis in crop plants. *Annu. Rev. Plant Biol.* **2013**, *64*, 71–88.

26. Kong, W.L.; Zhang, C.H.; Zhang, S.C.; Qiang, Y.L.; Zhang, Y.; Zhong, H.; Li, Y.S. Uncovering the novel QTLs and candidate genes of sale tolerance in rice with linkage mapping, RTM-GWAS, and RNA-seq. *Rice* **2021**, *14*, 93. [CrossRef]

27. Wuriyanghan, H.; Zhang, B.; Cao, W.H.; Ma, B.; Lei, G.; Liu, Y.F.; Wei, W.; Wu, H.J.; Chen, L.J.; Chen, H.W.; et al. The ethylene receptor ETR2 delays floral transition and affects starch accumulation in rice. *Plant Cell.* **2009**, *21*, 1473–1494. [CrossRef]

28. Yu, B.S.; Lin, Z.W.; Li, H.X.; Li, X.J.; Li, J.Y.; Wang, Y.H.; Zhang, X.; Zhu, Z.F.; Zhai, W.X.; Wang, X.K.; et al. TAC1, a major quantitative trait locus controlling tiller angle in rice. *Plant J.* **2007**, *52*, 891–898. [CrossRef]

29. Na, J.K.; Huh, S.M.; Yoon, I.S.; Byun, M.O.; Lee, Y.H.; Lee, K.O.; Kim, D. Rice LIM protein OsPLIM2a is involved in rice seed and tiller development. *Mol. Breed.* **2014**, *34*, 569–581. [CrossRef]

30. Das, S.; Parida, S.K.; Agarwal, P.; Tyagi, A.K. Transcription factor OsNF-YB9 regulates reproductive growth and development in rice. *Planta* **2019**, *250*, 1849–1865. [CrossRef]

31. Mathew, I.E.; Das, S.; Mahto, A.; Agarwal, P. Three rice NAC transcription factors heteromerize and are associated with seed size. *Front. Plant Sci.* **2016**, *7*, 1638. [CrossRef] [PubMed]

32. Wang, L.; Lu, Q.T.; Wen, X.G.; Lu, C.M. Enhanced Sucrose Loading Improves Rice Yield by Increasing Grain Size. *Plant Physiol.* **2015**, *169*, 2848–2862. [CrossRef] [PubMed]

33. Griffiths, C.A.; Sagar, R.; Geng, Y.; Primavesi, L.F.; Patel, M.K.; Passarelli, M.K.; Gilmore, I.S.; Steven, R.T.; Bunch, J.; Paul, M.J.; et al. Chemical intervention in plant sugar signalling increases yield and resilience. *Nature* **2016**, *540*, 574–578. [CrossRef] [PubMed]

34. Jung, H.; Lee, D.K.; Choi, Y.D.; Kim, J.K. OsIAA6, a member of the rice Aux/IAA gene family, is involved in drought tolerance and tiller outgrowth. *Plant Sci.* **2015**, *236*, 304–312. [CrossRef]

35. Huang, X.H.; Yang, S.H.; Gong, J.Y.; Zhao, Q.; Feng, Q.; Zhan, Q.L.; Zhao, Y.; Li, W.J.; Cheng, B.Y.; Xia, J.H. Genomic architecture of heterosis for yield traits in rice. *Nature* **2016**, *537*, 629–633. [CrossRef]

36. Li, D.Y.; Huang, Z.Y.; Song, S.H.; Xin, Y.Y.; Mao, D.H.; Lv, Q.M.; Zhou, M.; Tian, D.M.; Tang, M.F.; Wu, Q.; et al. Integrated analysis of phenome, genome, and transcriptome of hybrid rice uncovered multiple heterosis-related loci for yield increase. *Proc. Natl. Acad. Sci. USA* **2016**, *113*, E6026–E6035. [CrossRef]

37. Lin, Z.C.; Qin, P.; Zhang, X.W.; Fu, C.J.; Deng, H.C.; Fu, X.X.; Hunag, Z.; Jiang, S.Q.; Tang, X.Y.; Wang, X.F.; et al. Divergent selection and genetic introgression shape the genome landscape of heterosis in hybrid rice. *Proc. Natl. Acad. Sci. USA* **2020**, *117*, 4623–4631. [CrossRef]

38. Zhang, X.Y.; Jia, H.Y.; Li, T.; Wu, J.Z.; Nagarajan, R.; Lei, L.; Powers, C.; Kan, C.C.; Hua, W.; Liu, Z.Y.; et al. TaCol-B5 modifies spike architecture and enhances grain yield in wheat. *Science* **2022**, *376*, 180–183. [CrossRef]

39. Zhang, Z.Y.; Sun, X.M.; Ma, X.Q.; Xu, B.X.; Zhao, Y.; Ma, Z.Q.; Li, G.L.; Khan, N.U.; Pan, Y.H.; Liang, Y.T.; et al. GNP6, a novel allele of MOC1, regulates panicle and tiller development in rice. *Crop J.* **2020**, *9*, 57–67. [CrossRef]

Article

Natural Variation of *OsHd8* Regulates Heading Date in Rice

Huanran Yuan [1,†], Ruihua Wang [1,†], Mingxing Cheng [1,†], Xiao Wei [1], Wei Wang [1], Fengfeng Fan [1], Licheng Zhang [1], Zhikai Wang [2], Zhihong Tian [2] and Shaoqing Li [1,*]

1   State Key Laboratory of Hybrid Rice, Hongshan Laboratory of Hubei Province, Key Laboratory for Research and Utilization of Heterosis in Indica Rice of Ministry of Agriculture, Engineering Research Center for Plant Biotechnology and Germplasm Utilization of Ministry of Education, College of Life Science, Wuhan University, Wuhan 430072, China
2   College of Life Science, Yangtze University, Jingzhou 434025, China
*   Correspondence: shaoqingli@whu.edu.cn
†   These authors contributed equally to this work.

Abstract: Heading date, as one of the most important agronomic traits, is a fundamental factor determining crop yield. Although diverse genes related to heading date have already been reported in rice, the key gene that regulates heading date is still poorly understood. Here, we identified a heading date regulator, heading date 8 (*OsHd8*), which promoted the heading date under long-day conditions and encoded a putative HAP3 subunit of the CCAAT-box-binding transcription factor. It is localized in the nucleus and expressed in various tissues. Sequence analysis revealed that there were four SNPs and one InDel in the promoter region of *OsHd8*, which was involved in the regulation of some floral regulators including *GHD7.1*, *SDG718*, *OsGI* and *HDT1*. Further evolutionary analysis showed that *OsHd8* presents divergence between *indica* and *japonica*, showing natural selection during the domestication of cultivated rice. These results indicate that *OsHd8* plays an important role in the regulation of heading date, and may be an important target for rice breeding programs.

Keywords: heading date; *Heading date 8 (OsHd8)*; promoter; natural variation; rice

Citation: Yuan, H.; Wang, R.; Cheng, M.; Wei, X.; Wang, W.; Fan, F.; Zhang, L.; Wang, Z.; Tian, Z.; Li, S. Natural Variation of *OsHd8* Regulates Heading Date in Rice. *Agronomy* **2022**, *12*, 2260. https://doi.org/10.3390/agronomy12102260

Academic Editor: Argelia Lorence

Received: 22 August 2022
Accepted: 16 September 2022
Published: 21 September 2022

Publisher's Note: MDPI stays neutral with regard to jurisdictional claims in published maps and institutional affiliations.

1. Introduction

Flowering is an important process in plant transition from vegetative to reproductive growth [1,2]. Rice flowering is often affected by the external environment, including light, temperature, and nutritional conditions [3]. Appropriate flowering time is not only beneficial to the reproductive development of rice but also affects the yield of rice [4].

To date, many flowering-related genes have been identified in plants [5–10]. *Ghd7* encodes a protein with a CCT (CO, CO-LIKE and TIMING OF CAB1) domain that delays the heading date, increases the plant height and promotes panicle size development [5]. In *Arabidopsis thaliana*, *CO* (*CONSTANS*), as a key factor in the photoperiod pathway, promotes flowering under long-day (LD) conditions [11]. FT (FLOWERING LOCUS T) is a member of the *PEBP* gene family that shares homology with RAF kinase inhibitor proteins (RKIPs; these are activated by CO to regulate flowering [12–14]. *Hd1* and *Hd3a*, a rice ortholog of the *Arabidopsis CO* and *FT* gene, respectively, have been identified to promote heading under short day (SD) conditions [6,9]. However, the regulation of *Hd1* and *Hd3a* in rice is different from that in *Arabidopsis thaliana*. *PhyB* (*phytochromes B*) is involved in the post-translation regulation of *Hd1* to regulate *Hd3a* expression, while overexpression of *Hd1* inhibits *Hd3a* expression and delays flowering depending on phyB under SD conditions [15]. *RFT1* (*RICE FLOWERING LOCUS T 1*), encoding the mobile flowering signal, is the closest homolog of Hd3a. In *Hd3a*-RNAi transgenic plants, the *RFT1* gene is activated to promote flowering under SD conditions [10,16]. *Ehd1* (*Early heading date 1*) encodes a B-type response regulator, which can promote the flowering of rice under SD conditions by regulating expression of the *FT-like* gene [8]. *Ehd1*, as a unique floral activation pathway in rice, regulates the heading date by the protein complex of Ghd7-Hd1 and OsRE1-OsRIP1 [17–19]. PPS (Peter Pan syndrome) is a homolog of the photomorphogenic gene *COP1* (*CONSTITUTIVE PHOTOMORPHOGENIC1*) in *Arabidopsis thaliana* and promotes flowering by regulating the GA biosynthesis and suppressing the *miR156/miR172* expression [8]. In addition, MADS-box genes have been shown to play a vital role in the flowering time, of which, *OsMADS50* is an important flowering activator, controlling various downstream floral regulators in rice, including *OsMADS1*, *OsMADS4*, *OsMADS15*, *OsMADS18* and *Hd3a* [20–23].

The HAP (heterotrimeric heme activator protein) family, a class of CCAAT box factor (CBF) or nuclear factor Y (NF-Y), have been identified as important regulators in rice [24]. The HAP complex consists of three subunits, namely HAP2 (NF-YA; CBF-B), HAP3 (NF-YB; CBF-A) and HAP5 (NF-YC; CBF-C) [25,26]. A total of 10 *HAP2* (*OsHAP2A–J*) genes, 11 *HAP3* (*OsHAP3A–K*) genes and 7 *HAP5* (*OsHAP5A–G*) genes have been identified in rice, and recent studies revealed that *HAP* gene members play a vital role in the rice heading date [25]. HAP5B and HAP5D, as flowering inhibitory factors, directly interact with HAP3D, HAP3F and HAP3H protein to regulate the photoperiodic flowering response of rice under LD [27]. *DHD1* (*DELAYED HEADING DATE1*) is involved in the flowering development through binding to their target *HAP5* family genes *HAP5C* and *HAP5D* [28]. *OsHAPL1* (*Heme Activator Protein like 1*), as a flowering repressor, can physically interact with the DTH8/HAP3H and Hd1, and repress the heading date in rice [4].

Transcriptional regulation is largely controlled through gene promoters and their contributing cis-acting elements (CREs) [29]. The diversity of cis-regulatory elements including auxin response elements (AuxREs), the abscisic acid response element (ABRE), A-box (TACGTA), C-box (GACGTC), and G-box (CACGTG) are associated with auxin response in *Arabidopsis thaliana* [30]. Introducing new or disrupting existing upstream CREs, including single-nucleotide polymorphisms (SNPs), segmental deletions, insertion of transposable elements, and copy number variations, often leads to changes in many agronomic characters, resulting in crop improvement [31–34]. 3-bp InDel in the promoter region of *Sl-ALMT9* (*Al-ACTIVATED MALATE TRANSPORTER9*) disrupts a W-box binding site, which prevents binding of the transcription repressor WRKY42, and promotes the accumulation of malic acid in fruits [35]. *OsREM20* (*Oryza sativa REPRODUCTIVE MERISTEM 20*) encodes a B3 domain transcription factor and controls the grain number per panicle in rice by affecting the binding efficiency of *OsMADS34* to the CArG box in the promoter [33].

In this study, we identified an early-flowering gene *OsHd8* in the early-flowering rice JiaHong2B (JH2B) through map-based cloning, which encodes a putative HAP3 subunit of the CCAAT-box-binding transcription factor, and regulates the expression of *OsGI*, *OsSDG718*, *OsHDT1* and *OsGHD7.1* to promote rice heading under long-day conditions

2. Materials and Methods

2.1. Plant Materials and Cultivation

To clone *OsHd8*, a F_2 population of 4500 individuals was generated from the cross of 1880/JH2B. A set of germplasms, including 101 rice varieties (59 *indica* (IND), 9 *temperate japonica* (TEJ), 9 *tropical japonica* (TRJ), 12 *aus* (AUS), 2 *aromatic* (ARO) and 10 *admix* (ADM)), were used for genotyping of *OsHd8* (Supplementary Table S1). All the rice plants were grown in the ErZhou experimental field (30°40′ N, 114°88′ E) of Wuhan University.

2.2. Heading Date Investigation Map-Based Cloning

The 1880, JH2B two parents and F_2 population were investigated for the heading date. Each plant recorded the heading date when the rice plant began to head with a single spike. A total of 606 F_2 plants with extreme phenotypes, which defined as the first 15% of heading in the whole F_2 population, derived from the cross of 1880/JH2B were used for mapping of *OsHd8*. Firstly, it was primarily located in the 3.3~6.53 Mb region of the chromosome 8 by BSA-Seq (bulked segregant sequencing), and then we used the F_2 population of the target

QTL and mapped *OsHd8* between the marker SNP7730 and SSR-1 (Supplementary Table S2) using a chromosome fragment substitution analysis.

2.3. BSA-Seq and Analysis of the Seq-BSA Data

For the BSA-seq, two DNA pools were developed by selecting the extreme early heading date plants and extreme late heading date plants from the F_2 population. The extreme phenotype of the early growth period was defined as the first 15% of heading in the whole F_2 population, and the extreme phenotype of the late growth period was defined as the last 15% of the heading in the whole F_2 population. The early heading date pool (Z-pool) was made by mixing equal amounts of DNA from 50 extreme early heading date plats, and the late heading date pool (W-pool) was made by mixing equal amounts of DNA from 50 extreme late heading date plats. The DNA isolated from the two DNA pools were prepared for BSA sequencing.

Libraries for all the DNA pools were prepared according to the Illumina TruSeq Library Construction Kit. The DNA libraries were sequenced on Illumina Hiseq Xten PE15 (Illumina Inc., San Diego, CA, USA). The short reads from the two DNA pools were aligned to a *Nipponbare* reference genome (MSU Rice Genome Annotation Project Release 7) using the BWA software [36]. Reads of the Z-pool and W-pool were separately aligned to a *Nipponbare* reference genome and consensus sequence reads to call SNPs with the SAM tools software [36]. The SNP loci between the test samples and reference genome were obtained using the GATK software [37]. The Euclidean distance (ED) and SNP-index were calculated to identify the candidate regions of the genome associated with the heading date [38].

2.4. RNA Isolation and Quantitative Reverse Transcription-PCR (qRT-PCR) Analysis

Total RNAs were extracted from young rice leaves of 3-week-old seedlings using the TRIzol Reagent according to the manufacturer's protocol (Invitrogen, Carlsbad, CA, USA). Total RNA was used for synthesizing the cDNA with a reverse transcription kit (Vazyme, Nanjing, China). Quantitative reverse transcriptase (qRT-PCR) was performed on a Roche LightCycler®480II instrument using the Hieff qPCR SYBR Green Master Mix (No Rox) following the manufacturer's instructions (Yeasen, Shanghai, China). The actin gene was used as the internal control. All assays were performed with three biological replicates and the relative expression level was analyzed with the $2^{-\Delta\Delta CT}$ method [39].

2.5. Vector Constructions and Plant Transformation

In order to construct a complementary vector, the upstream 2 kb promoter region of the ATG, gene coding region, and downstream 1 kb region of the *OsHd8* gene, were amplified from the genomic DNA of 1880 and then constructed into the complementary vector pCAMBIA1301. To prepare the construction of the Hd8 overexpression vector, the CDS of *OsHd8* was amplified from 1880 and was introduced into the vector pCAMBIA1301-Ubi. All the constructed vector plasmids were transformed into an *Agrobacterium tumefactions* strain EHA105 and transferred by *Agrobacterium*-mediated transformation into JH2B.

2.6. Dual Luciferase (LUC) Analysis

The promoter regions were amplified from 1880 and JH2B, then cloned into the pGreenII 0800-LUC vector. Subsequently, the vectors co-transformed with GV3101 (pSoup-p19) chemically competent cells. Overnight, *A. tumefaciens* were cultured at 28 °C and collected by centrifugation and re-suspended in the MS medium with $OD_{600} = 1.5$, and incubated at 28 °C and 200 rpm for 3 h. The strains were infiltrated into tobacco (*Nicotiana benthamiana*) leaves and tested after 3 days (long day/white light). Leaves were infiltrated with 1 mM D-luciferin solution and images were captured using a Tanon 5200 Multi imaging system. Quantification was performed using the Dual-Luciferase Reporter Assay System (Promega, Madison, WI, USA). All the assays were performed on three biological replicates.

2.7. Evolutionary Analysis

The genome sequences of 3024 cultivars and 32 wild rice accessions were obtained from the Rice Functional Genomics and Breeding Database (RFGB, http://www.rmbreeding.cn/Snp3k, accessed on 20 April 2022) and OryzaGenome (http://viewer.shigen.info/oryzagenome/, accessed on 22 April 2022) [40,41]. The geographic information of cultivated rice populations was obtained from the MKBASE (http://www.mbkbase.org/rice/germplasm, accessed on 28 April 2022) and marked on the map using R software. The PopGenome package in the R software was used to calculate the parameters of genetic divergence for *OsHd8* and its flanking regions between *indica* and *japonica* subspecies, including haplotype and nucleotide F_{ST}, Nei's G_{ST}, Hudson's G_{ST} and H_{ST} [42]. A phylogenetic tree of *OsHd8* was constructed using the UPGMA method with MEGA7.0 [43], and a haplotype network was constructed using the pegas package in the R software [44].

2.8. Statistical Analysis

All the assays were performed on three biological replicates. The data were analyzed using the GraphPad Prism 9 software (https://www.graphpad.com/, accessed on 28 January 2022) and the means were compared by Student's *t*-test, the *, ** and *** mean $p < 0.05, 0.01$ and 0.001, respectively. The primers used for genetic mapping, vector construction, PCR and qRT-PCR analysis were all listed in Supplementary Table S4.

3. Results

3.1. Genetic Analysis and Mapping of OsHd8 for Heading Date

In the breeding practice, we found that the heading date of the *O. longistaminata* introgression line 1880 and an early flowering variety JiaHong2B (JH2B), was about 92 days and 65 days (Figure 1A,B), respectively, showing a great difference. To explore the genetic basis for the heading date in 1880, genetic linkage analysis of 1147 F_2 individuals derived from the cross of 1880/JH2B displayed a continuous distribution with an apparent valley bottom between 62 and 92 days (Figure 1C). These plants were then used for the short and long heading date pools. The two pools were then subjected to whole-genome sequencing up to >121× coverage, and 835,204 high-quality single-nucleotide polymorphisms (SNPs) were identified. SNP-index analysis showed that there was only one obvious single peak on the short arm 3.3~6.53 Mb of chromosome 8 (Supplementary Figure S1), meaning that the candidate gene controlling heading date is possibly located in this region, and named as *Hd8*.

A total of 660 plants with extreme phenotypes were then selected from a 4500 F_2 population of 1880/JH2B cross and were used for fine mapping; the *OsHd8* successfully narrowed the locus to a 31.8 kb region between the marker SNP7730 and SSR-1 (Figure 2A). According to the information from the RGAP (Rice Genome Annotation Project), four predicted genes were present in this region, namely, *LOC_Os08g07730*, *LOC_Os08g07740*, *LOC_Os08g07760*, and *LOC_Os08g07774* (Figure 2A). qRT-PCR showed that the *LOC_Os08g07740*, encoding a histone-like transcription factor and archaeal histone, had a large expressional difference between 1880 and JH2B (Figure 2B–E). And it was reported a flowering suppressor named *EF8/LHD1* [45,46], we deduced that the *LOC_Os08g07740* was responsible for *OsHd8*.

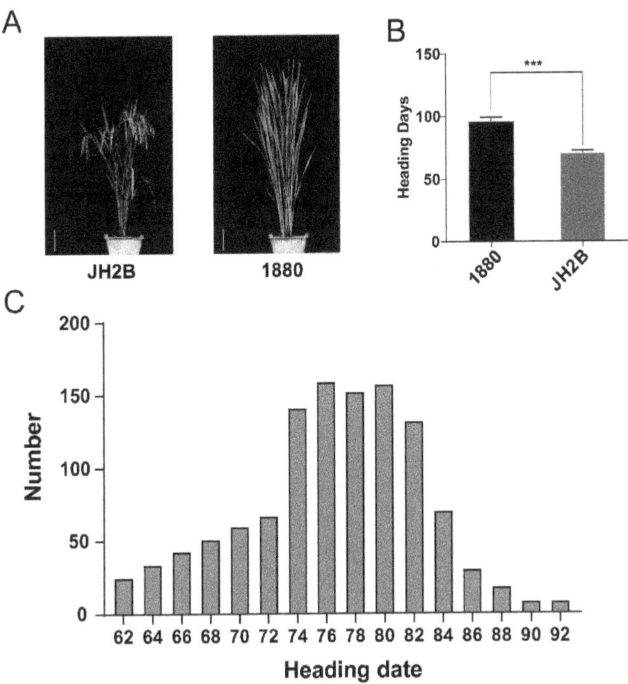

Figure 1. Genetic analysis of heading date in 1880 and JH2B. (**A**) The phenotype of the 1880 and JH2B for heading date. Scale bars = 5 cm; (**B**) Statistics on the heading date of 1880 and JH2B; Values are means $\pm$ SD (n = 5), *** indicates a significant difference at $p < 0.001$ by t-test;(**C**) Distribution of heading date in the F_2 population derived from 1880 $\times$ JH2B.

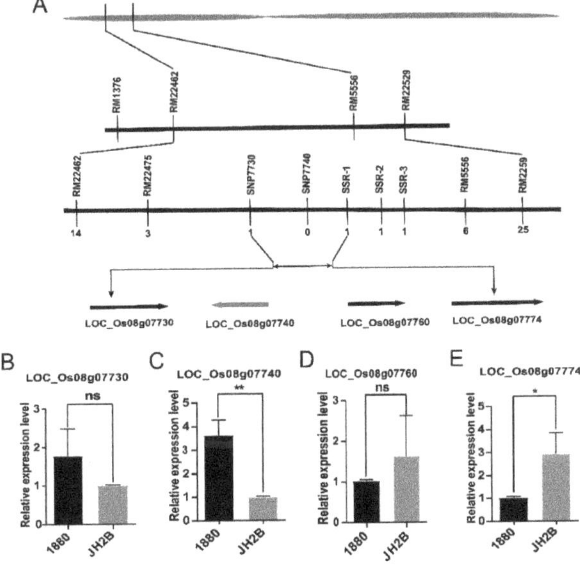

Figure 2. Identification of candidate heading date gene for *OsHd8*. (**A**) Fine mapping of *OsHd8*. The *OsHd8* locus was detected on chromosome 8. Positional cloning narrowed to a 31.8 kb region between

marker SNP7730 and SSR-1. The red colors marked gene *LOC_Os08g7740* was the candidate gene; (**B–E**) Comparison of the relative expression levels of four candidate genes among 1880 and JH2B. Values are means ± SEM (n = 3), The asterisks indicate significant differences (*, $p < 0.05$; **, $p < 0.01$; ns means no significance; Student's *t*-test).

3.2. OsHd8 Encodes a Transcriptional Repressor

To investigate whether *LOC_Os08g07740* was responsible for the phenotypic changes, a 4.7 kb genomic fragment, including a 2 kb upstream regulatory sequence, a 1.7 kb gene coding region, and a 1 kb downstream fragment of *OsHd8* from 1880, was cloned into the vector pCAMBIA1301, and introduced into JH2B, through *agrobacterium*-mediated transformation. Compared to JH2B, the heading date of transgenic complementary plants was delayed by about one week and showed an increased expression level (Figure 3A–C).

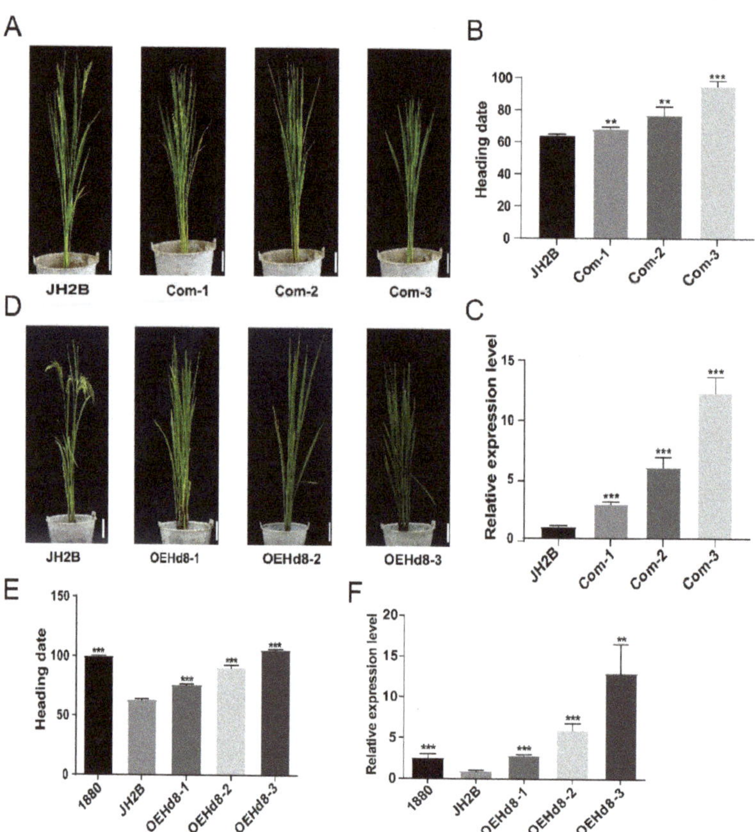

Figure 3. Effect of *OsHd8* on heading date in rice. (**A**) and (**D**) Comparison of heading date of JH2B, *OsHd8* complemented transgenic plants (Com-1-Com-3) and *OsHd8* overexpression transgenic plants (OEHd8-1-OEHd8-3). Scale bars = 5 cm; (**B**) and (**E**) Statistical analysis of heading date of JH2B, *OsHd8* complemented transgenic plants and *OsHd8* overexpression transgenic plants. Values are means ± SD (n = 5), The asterisks indicate significant differences (**, $p < 0.01$; ***, $p < 0.00$; Student's *t*-test); (**C**) and (**F**) Expression levels of *OsHd8* in the JH2B, *OsHd8* complemented transgenic plants and *OsHd8* overexpression transgenic plants. Values are means ± SEM (n = 3), The asterisks indicate significant differences (**, $p < 0.01$; ***, $p < 0.00$; Student's *t*-test).

To further validate the function of *LOC_Os08g07740*, an overexpression construct was transformed into JH2B. The *OsHd8* transcript level increased by about fivefold, and the heading date was delayed by 15 to 35 days compared with control plants (Figure 3D,E). These results demonstrated that *LOC_Os08g07740* is *OsHd8*, which is essential for regulating the heading date in rice.

To further investigate the function of *OsHd8*, protein subcellular localization was performed and found that it was located in the nucleus (Figure 4A); spatiotemporal analysis showed that *OsHd8* was expressed in all tissues, including the roots, young leaves, culms, and panicle (Figure 4B), corresponding to the A protein-BLAST(BLASTp) at NCBI online revealed that *OsHd8* encodes a CBFD_NFYB_HMF domain nuclear transcription factor belonging to the HAP3 subunit [46]. The transcriptional activity assays were then performed in rice protoplasts. The luciferase reporter gene contained five copies of binding sites for GAL4, and the Renilla luciferase gene was used as the internal reference. Compared with the transactivator control constructs GALBD-VP16, GALBD-VP16 fused with *OsHd8* induced significantly less LUC activity in rice protoplasts (Figure 4C,D), indicating that OsHd8 functions as a transcriptional repressor.

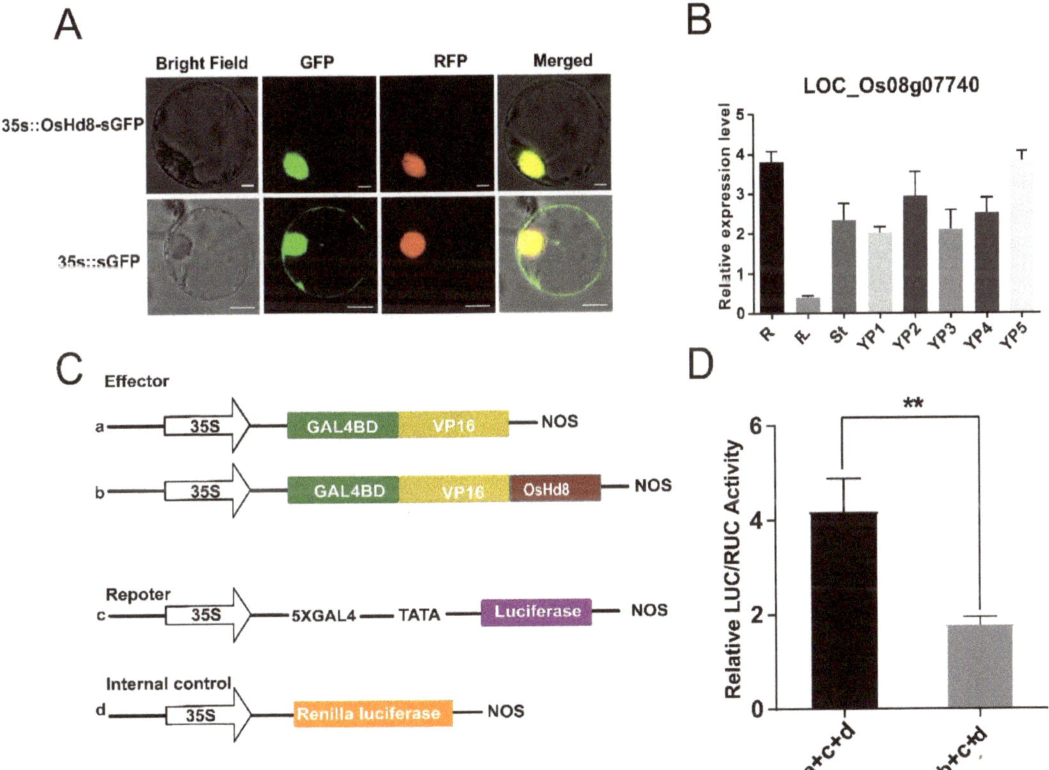

Figure 4. Molecular characterization of *OsHd8*. (**A**) Subcellular localization of OsHd8 (35S: OsHd8-sGFP) in rice protoplasts. Scale bar = 10 μm; (**B**) *OsHd8* expression levels in various organs revealed by qRT–PCR, including roots (R), Flag leaf (FL), stem node (St) and young panicles of different lengths (YP1: 0.5–1 cm young panicle, YP2: 1–2 cm young panicle, YP3: 2–3 cm young panicle, YP4: 3–4 cm young panicle, YP5: 4–5 cm young panicle). Values are means ± SEM (n = 3), ** indicates a significant difference at $p < 0.01$ by *t*-test; (**C**) The main structure of vectors of transcriptional activity assays; (**D**) The transcription activity in rice protoplasts by co-transformation of different effector vectors with the reporter plasmids and internal control vectors.

3.3. Expression Level of OsHd8 Affects Heading Date

To illustrate how *OsHd8* regulates the heading date, we compared the genomic sequence of *OsHd8* between 1880 and JH2B; no nucleotide difference was detected in the coding regions of *OsHd8* (Supplementary Figure S2). Instead, four polymorphisms and one 8-bp InDel were found in the promoters of the two parent lines (Figure 5A).

To verify whether the 8-bp InDel or 4 SNPs in the *OsHd8* promoter affect the *OsHd8* expression, we generated constructs by installing two type promoter fragments into the pGreenII0800-LUC vector and then introduced the constructs into rice protoplasts for transient expression assays. Results showed that 1880 promoter activity was much stronger than that of JH2 (Figure 5B). Furthermore, to validate the promoter activation capacity between 1880 and JH2B, a Dual-LUC assay was carried out in tobacco leaves and found that the 1880 promoter had a greater LUC/REN value than JH2B (Figure 5C), meaning that the sequence variations in the promoter of 1880 led to a high expression of *OsHd8*, and hence the delayed heading date.

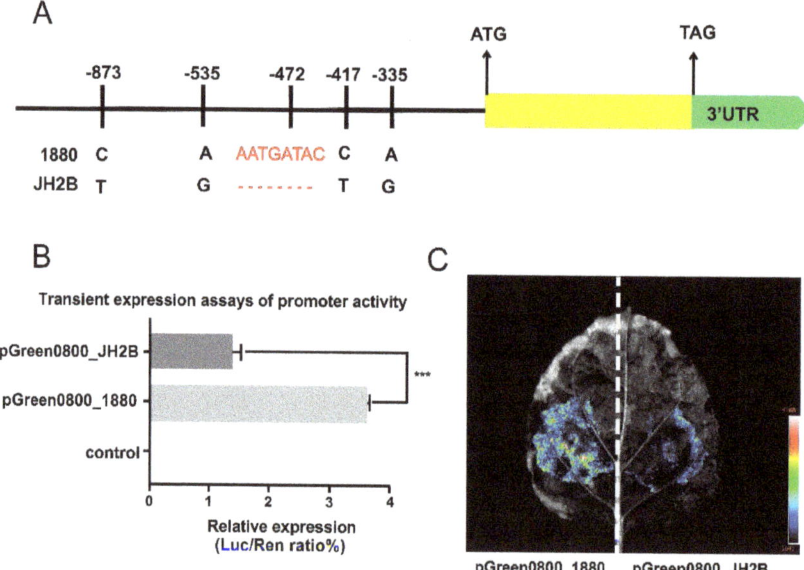

Figure 5. Sequence variations in the *OsHd8* promoter. (**A**) Schematic representation of the two major variations of the *OsHd8* promoter region. CT repeats are labeled in blue; red represents an InDel; (**B**) Transient expression assays of the two *OsHd8* promoter types in rice protoplasts. Values are means ± SEM (n = 6), *** indicates a significant difference at $p < 0.001$ by *t*-test; (**C**) Image of the Dual-LUC assay in tobacco leaves.

3.4. Comparative Transcriptome Analysis of JH2B and OE-OsHd8

To further explore the regulatory network underlying the *OsHd8* function, we performed RNA-sequencing analysis with young leaves from JH2B and OE-*Hd8*. A total of 1198 differentially expressed genes (DEGs) were identified, of which 825 genes were upregulated and 373 genes were downregulated in the OE-*OsHd8* plants. Gene Ontology (GO) assay showed that these DEGs were significantly enriched in terms of the metabolic process, binding and transcription regulator activity (Figure 6A). Further analysis of transcription factors of DEGs revealed that transcription factors associated with flowering were enriched in the *OsHd8* pathway, such as the MADS family, HAP2 family, bZIP family and WRKY family (Supplementary Figure S3) [25,47–49]. In particular, several genes controlling the rice heading date, such as *GHD7.1*, *SDG718*, *OsGI* and *HDT1* [2,50–52], were differentially expressed in OE-OsHd8 plants just as confirmed with qRT–PCR analysis (Figure 6B). These results demonstrated that *OsHd8* could be involved in complicated transcriptional regulation processes governing the rice heading date.

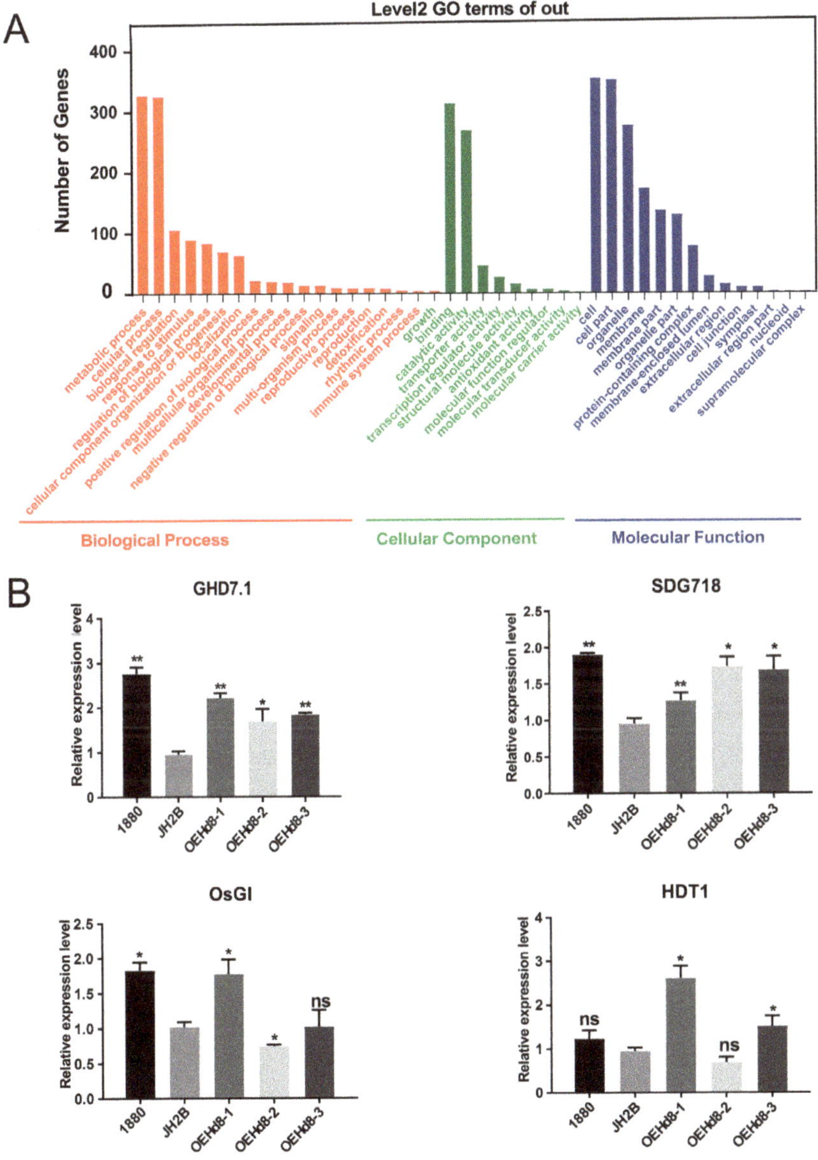

Figure 6. Transcriptome analysis of the DEGs from RNA-sequencing data in heading date between JH2B and OE-OsHd8. (**A**) Gene ontology (GO) annotation of differentially expressed genes (DEG) in heading date between JH2B and OE-OsHd8. The enriched GO terms of molecular function, biological process and cellular component were listed, and ranged from large to small according to -$\log_{10}^{p\text{value}}$; (**B**) Comparison of transcriptional expression levels of OsHd8 related genes in heading date between JH2B and OE-OsHd8. Values are means ± SEM (n = 3), The asterisks indicate significant differences (*, $p < 0.05$; **, $p < 0.01$; ns means no significance; Student's *t*-test).

3.5. Variations in the OsHd8 Promoter Affect its Expression and Heading Date

In order to understand the effects of the promoter variation on the expression of *OsHd8*, we then selected 101 rice accessions, including 59 *indica* (*IND*), 9 *temperate japonica* (*TEJ*), 9 *tropical japonica* (*TRJ*), 12 *aus* (*AUS*), 2 *aromatic* (*ARO*) and 10 *admix* (*ADM*) lines from different countries for clustering analysis (Figure 7A). Results showed that *OsHd8* diverges into six haplotypes (Figure 7A), of which the haplotype 2 (Hap2), belonging to JH2B, showed the shortest heading date. 1880 belongs to Hap1, which showed the longest heading date, while Hap3, Hap4 and Hap5 were derived from Hap1 by single-base mutations and had a heading date between the Hap1 and Hap2 (Figure 7B). Further transcriptional analysis of *OsHd8* in rice leaves showed that the Hap2 expression was significantly lower than the other five haplotypes (Figure 7C), consistent with the phenotype of the haplotypes. These results indicate that the promoter sequence variation seems significantly correlated with the gene expression and heading date of rice.

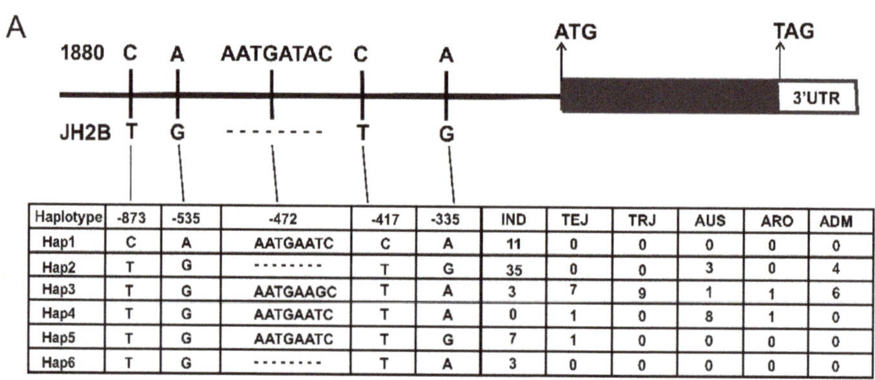

Haplotype	-873	-535	-472	-417	-335	IND	TEJ	TRJ	AUS	ARO	ADM
Hap1	C	A	AATGAATC	C	A	11	0	0	0	0	0
Hap2	T	G	- - - - - - - -	T	G	35	0	0	3	0	4
Hap3	T	G	AATGAAGC	T	A	3	7	9	1	1	6
Hap4	T	G	AATGAATC	T	A	0	1	0	8	1	0
Hap5	T	G	AATGAATC	T	G	7	1	0	0	0	0
Hap6	T	G	- - - - - - - -	T	A	3	0	0	0	0	0

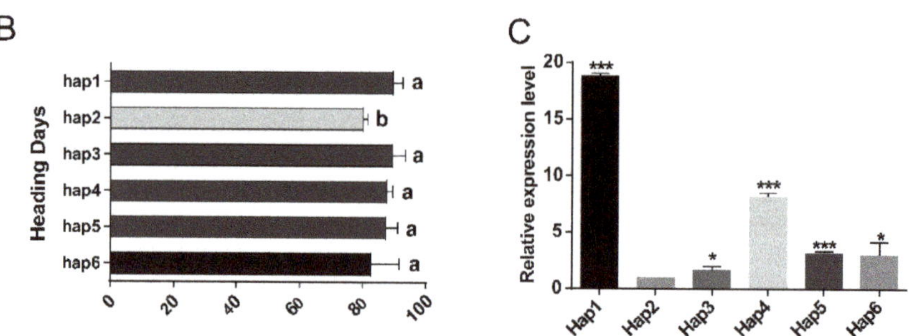

Figure 7. Haplotype analysis of *OsHd8*. (**A**) Haplotype analysis of *OsHd8* from 101 rice varieties; *IND* means *indica*, *TEJ* means *temperate japonica*, *TRJ* means *tropical japonica*, *AUS* means *aus*, *ARO* means *aromatic*, *ADM* means *admixed*; (**B**) Analysis of heading date in representative varieties with six haplotypes. n = 11 in Hap1, n = 42 in Hap2, n = 27 in Hap3, n = 10 in Hap4, n = 8 in Hap5, and n = 3 in Hap6; (**C**) The relative expression level of the six haplotypes. Values are means ± SEM (n = 3), The asterisks indicate significant differences (*, $p < 0.05$; ***, $p < 0.001$; Student's *t*-test).

3.6. OsHd8 Is Subjected to Selection in Cultivated Rice

To investigate the genetic relationship of *OsHd8* variations, a total of 3024 cultivated rice from the 3k database (http://www.rmbreeding.cn/Index/, accessed on 20 April 2022) and 32 *O. rufipogon* accessions (http://viewer.shigen.info/oryzagenome/, accessed on 20 April 2022) were selected to analyze the genetic diversity of this gene and its flanking region. The nucleotide diversity value (π) of *OsHd8* is lower than its flanking regions in both cultivated and *O. rufipogon* accessions (Figure 8A, Supplementary Table S3), suggesting that *OsHd8* might be subjected to natural selection.

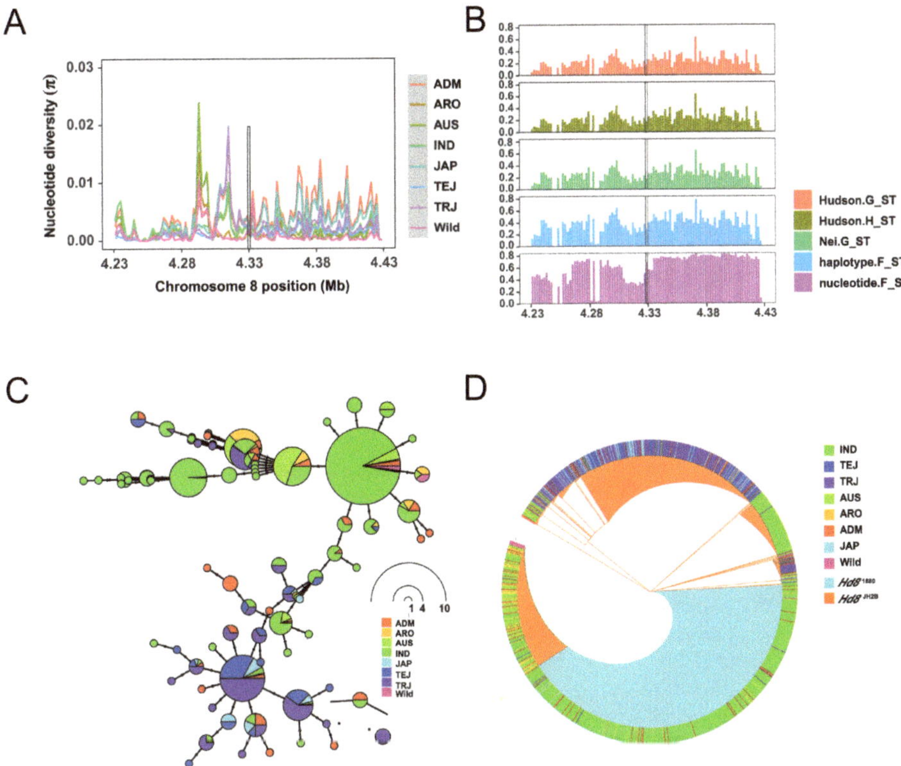

Figure 8. Phylogenetic relationship and genetic variation analysis of *OsHd8*. (**A**) Nucleotide diversity and selection analysis of *OsHd8* and its flanking region (~200 kb). From Rice SNPS-Seek Database (https://snp-seek.irri.org/_snp.zul, accessed on 20 April 2022) and OryzaGenome (http://viewer.shigen.info/oryzagenome/, accessed on 22 April 2022), Genomic sequences of 3024 cultivated germplasm and 32 wild germplasm were obtained. *ADM* means *admixed*, *ARO* means *aromatic*, *AUS* means *aus*, *IND* means *indica*, *JAP* means unclassified *japonica*, *TEJ* means *temperate japonica*, *TRJ* means *tropical japonica*; (**B**) The parameters of genetic difference between *indica* and *japonica* ecotypes for *OsHd8* and its flanking genomic regions; (**C**) Haplotype network of *OsHd8*. The size of the circle is proportional to the number of samples given to the order. The black dots on the lines indicate the mutation steps between haplotypes; (**D**) Phylogenetic relationship of *OsHd8* generated between 3024 cultivated rice and 32 wild rice accessions. The orange and blue rectangles represent the types of *Hd8*^{1880} and *Hd8*^{JH2B}, respectively.

We further analyzed the parameters of genetic divergence in the *OsHd8* locus between *indica* and *japonica* subspecies from 3024 cultivated accessions, including the estimates of haplotype and nucleotide F_{ST}, Nei's G_{ST}, and Hudson's G_{ST} and H_{ST}. Genetic analysis showed that the five parameters in the *OsHd8* locus were all greater than 0.25 between *indica* and *japonica* subspecies, and the haplotype and nucleotide F_{ST} reached to 0.549 and 0.764, respectively, (Figure 8B), suggesting strong genetic differentiation between *indica* and *japonica* subspecies at the *OsHd8* locus [53].

Phylogenetic analysis with the coding sequence indicated that the *OsHd8* could be categorized into 86 haplotypes, and that the *indica* rice was closer to the wild rice haplotypes (Figure 8C). Meanwhile, 1880 was clustered together with the wild rice, meaning *Hd8*^{1880} evolved from wild rice *O. rufipogon* (Figure 8D). These results suggest that *Hd8*^{1880}

originated from wild rice and that at least one mutational event was involved in the origin of $Hd8^{1880}$ in *indica* rice.

4. Discussion

Flowering is an important trait of plants, and the appropriate flowering time is responsible for the growth and successful sexual reproduction in flowering plants [45]. How to accurately control the flowering time in rice is of great practical significance for the improvement of rice yield. In the present study, we found an elite allele of *EF8/LHD1/DTH8/Ghd8* from indica 1880, which delays the heading date by about 27 days, compared to the JH2B (Figure 1) [45,46,54,55]. Previously, the functional *DTH8* allele from *cv. Asominori* delays the heading date by about 13 days compared to CSSL61 (1-bp deletion in the exon that carries the nonfunctional DTH8) [55]. In our study, the newly identified mutations in the promoter reduced the expression level of *OsHd8* in JH2B, leading to a shorter heading date (Figure 3). Furthermore, the *DTH8* allele from *japonica cv. Asominori* could down-regulate the transcription of *Ehd1* and *Hd3a* to regulate the heading date, and another *EF8* functional allele delays the heading date by regulating the *Hd3a* and *RFT1*. However, our results indicated that *EF8* could alter the expression patterns of the other genes, *GHD7.1*, *SDG718*, *OsGI* and *HDT1* (Figure 3). Therefore, we speculate that different alleles of a gene may have different regulatory roles and target genes, resulting in different regulatory mechanisms and phenotypes. Whether differences in gene sequences or genetic backgrounds lead to different allelic effects is an interesting question that deserves further investigation.

The promoter is located in the upstream of the gene coding region and contains many cis-acting elements (CRE), and transcriptional regulation is mainly determined by the promoter CREs [29]. In our study, the promoter activity of 1880 is much higher than that of JH2B (Figure 4B,C) because of the SNP and InDel variations in the JH2B promoter (Figure 4A), of which one SNP at -335 bp (G/A) is located in the ABRE cis-acting element in the JH2B promoter (Figure 7). It has been well characterized that ABRE, as an important response element of abscisic acid (ABA), plays an important role in the regulation of the ABA signal network [56]. When encountered with drought stress, plants may accelerate the initiation of the flowering transformation to shorten their growth cycle through the RCN1 mediated ABA signal process [57]. Interestingly, we found that the ABA response element in the *OsHd8* promoter of the *japonica* variety *Nipponbare* with a short heading date was the same as that of JH2B. Haplotype analysis based on the promoter sequence showed that the JH2B haplotype had the shortest heading date and the lowest expression level (Figure 7). This reminds us that the mutation of the ABA-responsive element of *OsHd8* in JH2B may lead to the change in the growth period of rice. However, how ABA response elements regulate the expression of *OsHd8* needs to be further validated.

During the evolution of rice, the regional adaptability of cultivated rice is affected by the response to the length of daylight [58]. The heading date is determined by a variety of internal and external signals, including light time, temperature, and hormones [59]. The difference in sensitivity of rice to the photocycle and temperature makes the rice vary greatly in different areas. Many rice lines and wild rice in tropical and subtropical regions, such as *O. rufipogon*, have strong photoperiod sensitivity to flowering. This strong photoperiod sensitivity completely inhibits heading in long-day conditions, allowing it to induce heading only in short daylight conditions [60]. The $Hd8^{JH2B}$ allele cloned in this study may be a potential genetic resource. This genotype promotes flowering in long-day conditions and is widely present in the 3K database (Figure 8). With further analysis of the regional distribution map of $Hd8^{1880}$ and $Hd8^{JH2B}$ haplotypes, we can see that the $Hd8^{JH2B}$ haplotype distributes mainly in the higher latitude area (Figure 9), meaning that this haplotype rice may be less sensitive to photoperiod and have expanded more greatly than the other haplotype rice.

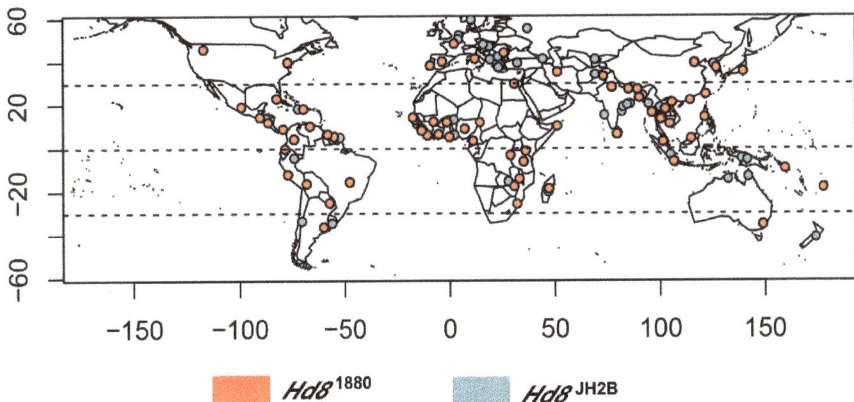

Figure 9. Geographical distribution of $Hd8^{1880}$ and $Hd8^{JH2B}$ haplotypes. The orange dots represent the geographical distribution of the haplotype $Hd8^{1880}$, and the cyan dots represent the geographical distribution of the haplotype $Hd8^{JH2B}$.

5. Conclusions

This study identified and mapped a heading date gene, *OsHd8*, from an early-flowering rice JiaHong2B (JH2B). Four SNPs and one InDel in the promoter region of *OsHd8* led to the advance of the JH2B heading date. Comparative transcriptome analysis revealed *OsHd8* to be involved in regulation of some floral regulators including *GHD7.1*, *SDG718*, *OsGI* and *HDT1*. *OsHd8* presents strong genetic differentiation between *indica* and *japonica* subspecies and shows artificial selection during the domestication of cultivated rice. Our work will provide a valuable heading date gene for rice breeding programs

Supplementary Materials: The supplementary figures and tables are available online at https://www.mdpi.com/article/10.3390/agronomy12102260/s1. Supplementary Figure S1: ED association analysis of candidate genes in heading date. Supplementary Figure S2: Comparison of CDS sequences of *OsHd8* gene in 1880, JH2B and *Nipponbare*. Supplementary Figure S3: Analysis of transcription factors belonging to differentially expressed genes in JH2B and OE-*OsHd8* expression profile by MAPMAN software. Supplementary Table S1: *OsHd8* haplotypes in the 101 rice cultivars. Supplementary Table S2: Recombination events between OsHd8 and molecular markers. Supplementary Table S3: The estimated parameters of nucleotide diversity (π) of *OsHd8* and its flanking regions. Supplementary Table S4: Primers used in this study.

Author Contributions: Conceptualization, S.L. and H.Y.; methodology, S.L.; software, M.C.; validation, R.W.; formal analysis, H.Y. and R.W.; investigation, X.W. and W.W.; resources, S.L. and F.F.; data curation, L.Z. and Z.W.; writing—original draft preparation, H.Y., R.W. and M.C.; writing—review and editing, H.Y. and S.L.; visualization, Z.T.; supervision, S.L.; project administration, S.L.; funding acquisition, S.L. All authors have read and agreed to the published version of the manuscript.

Funding: This work was supported by the National Natural Science Foundation of China (U20A2023, 31870322), Hubei Hongshan Laboratory (2021hszd010).

Institutional Review Board Statement: Not applicable.

Informed Consent Statement: Not applicable.

Data Availability Statement: The datasets generated during the current study are available from the corresponding author on reasonable request.

Acknowledgments: We thank IRRI for kindly providing the rice cultivars.

Conflicts of Interest: The authors declare no conflict of interest.

References

1. Simpson, G.G.; Dean, C. Arabidopsis, the Rosetta stone of flowering time? *Science* **2002**, *296*, 285–289. [CrossRef] [PubMed]
2. Hayama, R.; Yokoi, S.; Tamaki, S.; Yano, M.; Shimamoto, K. Adaptation of photoperiodic control pathways produces short-day flowering in rice. *Nature* **2003**, *422*, 719–722. [CrossRef] [PubMed]
3. Zhou, S.; Zhu, S.; Cui, S.; Hou, H.; Wu, H.; Hao, B.; Cai, L.; Xu, Z.; Liu, L.; Jiang, L.; et al. Transcriptional and post-transcriptional regulation of heading date in rice. *New Phytol.* **2021**, *230*, 943–956. [CrossRef]
4. Zhu, S.; Wang, J.; Cai, M.; Zhang, H.; Wu, F.; Xu, Y.; Li, C.; Cheng, Z.; Zhang, X.; Guo, X.; et al. The OsHAPL1-DTH8-Hd1 complex functions as the transcription regulator to repress heading date in rice. *J. Exp. Bot.* **2017**, *68*, 553–568. [CrossRef] [PubMed]
5. Xue, W.; Xing, Y.; Weng, X.; Zhao, Y.; Tang, W.; Wang, L.; Zhou, H.; Yu, S.; Xu, C.; Li, X.; et al. Natural variation in Ghd7 is an important regulator of heading date and yield potential in rice. *Nat. Genet.* **2008**, *40*, 761–767. [CrossRef]
6. Kojima, S.; Takahashi, Y.; Kobayashi, Y.; Monna, L.; Sasaki, T.; Araki, T.; Yano, M. Hd3a, a rice ortholog of the Arabidopsis FT gene, promotes transition to flowering downstream of Hd1 under short-day conditions. *Plant Cell Physiol.* **2002**, *43*, 1096–1105. [CrossRef]
7. Tanaka, N.; Itoh, H.; Sentoku, N.; Kojima, M.; Sakakibara, H.; Izawa, T.; Itoh, J.; Nagato, Y. The COP1 ortholog PPS regulates the juvenile-adult and vegetative-reproductive phase changes in rice. *Plant Cell* **2011**, *23*, 2143–2154. [CrossRef]
8. Doi, K.; Izawa, T.; Fuse, T.; Yamanouchi, U.; Kubo, T.; Shimatani, Z.; Yano, M.; Yoshimura, A. Ehd1, a B-type response regulator in rice, confers short-day promotion of flowering and controls FT-like gene expression independently of Hd1. *Genes Dev.* **2004**, *18*, 926–936. [CrossRef]
9. Yano, M.; Katayose, Y.; Ashikari, M.; Yamanouchi, U.; Monna, L.; Fuse, T.; Baba, T.; Yamamoto, K.; Umehara, Y.; Nagamura, Y. Hd1, a major photoperiod sensitivity quantitative trait locus in rice, is closely related to the Arabidopsis flowering time gene CONSTANS. *Plant Cell* **2000**, *12*, 2473–2483. [CrossRef]
10. Komiya, R.; Yokoi, S.; Shimamoto, K. A gene network for long-day flowering activates RFT1 encoding a mobile flowering signal in rice. *Development* **2009**, *136*, 3443–3450. [CrossRef]
11. Putterill, J.; Robson, F.; Lee, K.; Simon, R.; Coupland, G. The CONSTANS gene of Arabidopsis promotes flowering and encodes a protein showing similarities to zinc finger transcription factors. *Cell* **1995**, *80*, 847–857. [CrossRef]
12. Jin, S.; Nasim, Z.; Susila, H.; Ahn, J.H. Evolution and functional diversification of FLOWERING LOCUS T/TERMINAL FLOWER 1 family genes in plants. *Semin. Cell Dev. Biol.* **2021**, *109*, 20–30. [CrossRef] [PubMed]
13. Nakamura, Y.; Lin, Y.C.; Watanabe, S.; Liu, Y.C.; Katsuyama, K.; Kanehara, K.; Inaba, K. High-Resolution Crystal Structure of Arabidopsis FLOWERING LOCUS T Illuminates Its Phospholipid-Binding Site in Flowering. *iScience* **2019**, *21*, 577–586. [CrossRef] [PubMed]
14. Corbesier, L.; Vincent, C.; Jang, S.; Fornara, F.; Fan, Q.; Searle, I.; Giakountis, A.; Farrona, S.; Gissot, L.; Turnbull, C.; et al. FT protein movement contributes to long-distance signaling in floral induction of Arabidopsis. *Science* **2007**, *316*, 1030–1033. [CrossRef]
15. Ishikawa, R.; Aoki, M.; Kurotani, K.-i.; Yokoi, S.; Shinomura, T.; Takano, M.; Shimamoto, K. Phytochrome B regulates Heading date 1 (Hd1)-mediated expression of rice florigen Hd3a and critical day length in rice. *Mol. Genet. Genom.* **2011**, *285*, 461–470. [CrossRef]
16. Komiya, R.; Ikegami, A.; Tamaki, S.; Yokoi, S.; Shimamoto, K. Hd3a and RFT1 are essential for flowering in rice. *Development* **2008**, *135*, 767–774. [CrossRef]
17. Chai, J.; Zhu, S.; Li, C.; Wang, C.; Cai, M.; Zheng, X.; Zhou, L.; Zhang, H.; Sheng, P.; Wu, M.; et al. OsRE1 interacts with OsRIP1 to regulate rice heading date by finely modulating Ehd1 expression. *Plant Biotechnol. J.* **2021**, *19*, 300–310. [CrossRef]
18. Nemoto, Y.; Nonoue, Y.; Yano, M.; Izawa, T. Hd1, a CONSTANS ortholog in rice, functions as an Ehd1 repressor through interaction with monocot-specific CCT-domain protein Ghd7. *Plant J.* **2016**, *86*, 221–233. [CrossRef]
19. Shrestha, R.; Gomez-Ariza, J.; Brambilla, V.; Fornara, F. Molecular control of seasonal flowering in rice, arabidopsis and temperate cereals. *Ann. Bot.* **2014**, *114*, 1445–1458. [CrossRef]
20. Lee, S.; Kim, J.; Han, J.J.; Han, M.J.; An, G. Functional analyses of the flowering time gene OsMADS50, the putative SUPPRESSOR OF OVEREXPRESSION OF CO 1/AGAMOUS-LIKE 20 (SOC1/AGL20) ortholog in rice. *Plant J.* **2004**, *38*, 754–764. [CrossRef]
21. Zhao, J.; Chen, H.; Ren, D.; Tang, H.; Qiu, R.; Feng, J.; Long, Y.; Niu, B.; Chen, D.; Zhong, T.; et al. Genetic interactions between diverged alleles of Early heading date 1 (Ehd1) and Heading date 3a (Hd3a)/RICE FLOWERING LOCUS T1 (RFT1) control differential heading and contribute to regional adaptation in rice (*Oryza sativa*). *New Phytol.* **2015**, *208*, 936–948. [CrossRef] [PubMed]
22. Lu, S.-J.; Wei, H.; Wang, Y.; Wang, H.-M.; Yang, R.-F.; Zhang, X.-B.; Tu, J.-M. Overexpression of a Transcription Factor OsMADS15 Modifies Plant Architecture and Flowering Time in Rice (*Oryza sativa* L.). *Plant Mol. Biol. Report.* **2012**, *30*, 1461–1469. [CrossRef]
23. Kobayashi, K.; Yasuno, N.; Sato, Y.; Yoda, M.; Yamazaki, R.; Kimizu, M.; Yoshida, H.; Nagamura, Y.; Kyozuka, J. Inflorescence meristem identity in rice is specified by overlapping functions of three AP1/FUL-like MADS box genes and PAP2, a SEPALLATA MADS box gene. *Plant Cell* **2012**, *24*, 1848–1859. [CrossRef]
24. Kusnetsov, V.; Landsberger, M.; Meurer, J.; Oelmuller, R. The assembly of the CAAT-box binding complex at a photosynthesis gene promoter is regulated by light, cytokinin, and the stage of the plastids. *J. Biol. Chem.* **1999**, *274*, 36009–36014. [CrossRef]
25. Li, Q.; Yan, W.; Chen, H.; Tan, C.; Han, Z.; Yao, W.; Li, G.; Yuan, M.; Xing, Y. Duplication of OsHAP family genes and their association with heading date in rice. *J. Exp. Bot.* **2016**, *67*, 1759–1768. [CrossRef] [PubMed]
26. Thirumurugan, T.; Ito, Y.; Kubo, T.; Serizawa, A.; Kurata, N. Identification, characterization and interaction of HAP family genes in rice. *Mol. Genet. Genom. MGG* **2008**, *279*, 279–289. [CrossRef] [PubMed]

27. Kim, S.K.; Park, H.Y.; Jang, Y.H.; Lee, K.C.; Chung, Y.S.; Lee, J.H.; Kim, J.K. OsNF-YC2 and OsNF-YC4 proteins inhibit flowering under long-day conditions in rice. *Planta* **2016**, *243*, 563–576. [CrossRef]
28. Zhang, H.; Zhu, S.; Liu, T.; Wang, C.; Cheng, Z.; Zhang, X.; Chen, L.; Sheng, P.; Cai, M.; Li, C.; et al. DELAYED HEADING DATE1 interacts with OsHAP5C/D, delays flowering time and enhances yield in rice. *Plant Biotechnol. J.* **2019**, *17*, 531–539. [CrossRef]
29. Zou, C.; Sun, K.; Mackaluso, J.D.; Seddon, A.E.; Jin, R.; Thomashow, M.F.; Shiu, S.-H. Cis-regulatory code of stress-responsive transcription in *Arabidopsis thaliana*. *Proc. Natl. Acad. Sci. USA* **2011**, *108*, 14992–14997. [CrossRef]
30. Cherenkov, P.; Novikova, D.; Omelyanchuk, N.; Levitsky, V.; Grosse, I.; Weijers, D.; Mironova, V. Diversity of cis-regulatory elements associated with auxin response in *Arabidopsis thaliana*. *J. Exp. Bot.* **2018**, *69*, 329–339. [CrossRef]
31. Springer, N.; de Leon, N.; Grotewold, E. Challenges of Translating Gene Regulatory Information into Agronomic Improvements. *Trends Plant Sci.* **2019**, *24*, 1075–1082. [CrossRef] [PubMed]
32. Swinnen, G.; Goossens, A.; Pauwels, L. Lessons from Domestication: Targeting Cis-Regulatory Elements for Crop Improvement. *Trends Plant Sci.* **2016**, *21*, 506–515. [CrossRef]
33. Wu, X.; Liang, Y.; Gao, H.; Wang, J.; Zhao, Y.; Hua, L.; Yuan, Y.; Wang, A.; Zhang, X.; Liu, J.; et al. Enhancing rice grain production by manipulating the naturally evolved cis-regulatory element-containing inverted repeat sequence of OsREM20. *Mol. Plant* **2021**, *14*, 997–1011. [CrossRef]
34. Bai, X.; Huang, Y.; Hu, Y.; Liu, H.; Zhang, B.; Smaczniak, C.; Hu, G.; Han, Z.; Xing, Y. Duplication of an upstream silencer of FZP increases grain yield in rice. *Nat. Plants* **2017**, *3*, 885–893. [CrossRef]
35. Ye, J.; Wang, X.; Hu, T.; Zhang, F.; Wang, B.; Li, C.; Yang, T.; Li, H.; Lu, Y.; Giovannoni, J.J.; et al. An InDel in the Promoter of Al-ACTIVATED MALATE TRANSPORTER9 Selected during Tomato Domestication Determines Fruit Malate Contents and Aluminum Tolerance. *Plant Cell* **2017**, *29*, 2249–2268. [CrossRef]
36. Li, H.; Durbin, R. Fast and accurate short read alignment with Burrows-Wheeler transform. *Bioinformatics* **2009**, *25*, 1754–1760. [CrossRef]
37. McKenna, A.; Hanna, M.; Banks, E.; Sivachenko, A.; Cibulskis, K.; Kernytsky, A.; Garimella, K.; Altshuler, D.; Gabriel, S.; Daly, M.; et al. The Genome Analysis Toolkit: A MapReduce framework for analyzing next-generation DNA sequencing data. *Genome Res.* **2010**, *20*, 1297–1303. [CrossRef]
38. Hill, J.T.; Demarest, B.L.; Bisgrove, B.W.; Gorsi, B.; Su, Y.C.; Yost, H.J. MMAPPR: Mutation mapping analysis pipeline for pooled RNA-seq. *Genome Res.* **2013**, *23*, 687–697. [CrossRef]
39. Livak, K.J.; Schmittgen, T.D. Analysis of relative gene expression data using real-time quantitative PCR and the 2(-Delta Delta C(T)) Method. *Methods* **2001**, *25*, 402–408. [CrossRef]
40. Wang, C.C.; Yu, H.; Huang, J.; Wang, W.S.; Faruquee, M.; Zhang, F.; Zhao, X.Q.; Fu, B.Y.; Chen, K.; Zhang, H.L.; et al. Towards a deeper haplotype mining of complex traits in rice with RFGB v2.0. *Plant Biotechnol. J.* **2019**, *18*, 14–16. [CrossRef]
41. Ohyanagi, H.; Obayashi, T.; Yano, K. Editorial: Plant and Cell Physiology's 2016 Online Database Issue. *Plant Cell Physiol.* **2016**, *57*, 1–3. [CrossRef]
42. Pfeifer, B.; Wittelsburger, U.; Ramos-Onsins, S.E.; Lercher, M.J. PopGenome: An efficient Swiss army knife for population genomic analyses in R. *Mol. Biol. Evol.* **2014**, *31*, 1929–1936. [CrossRef]
43. Kumar, S.; Stecher, G.; Tamura, K. MEGA7: Molecular Evolutionary Genetics Analysis Version 7.0 for Bigger Datasets. *Mol. Biol. Evol.* **2016**, *33*, 1870–1874. [CrossRef]
44. Paradis, E. pegas: An R package for population genetics with an integrated-modular approach. *Bioinformatics* **2010**, *26*, 419–420. [CrossRef]
45. Feng, Z.; Zhang, L.; Yang, C.; Wu, T.; Lv, J.; Chen, Y.; Liu, X.; Liu, S.; Jiang, L.; Wan, J. EF8 is involved in photoperiodic flowering pathway and chlorophyll biogenesis in rice. *Plant Cell Rep.* **2014**, *33*, 2003–2014. [CrossRef] [PubMed]
46. Dai, X.; Ding, Y.; Tan, L.; Fu, Y.; Liu, F.; Zhu, Z.; Sun, X.; Sun, X.; Gu, P.; Cai, H.; et al. LHD1, an allele of DTH8/Ghd8, controls late heading date in common wild rice (Oryza rufipogon). *J. Integr. Plant Biol.* **2012**, *54*, 790–799. [CrossRef]
47. Hong-Gyu, K.; Seonghoe, J.; Jae-Eun, C.; Yong-Gu, C.; Gynbeung, A. Characterization of Two Rice MADS-Box Genes That Control Flowering Time. *Mol. Cells* **1997**, *7*, 559–566.
48. Li, X.; Tian, X.; He, M.; Liu, X.; Li, Z.; Tang, J.; Mei, E.; Xu, M.; Liu, Y.; Wang, Z.; et al. bZIP71 delays flowering by suppressing Ehd1 expression in rice. *J. Integr. Plant Biol.* **2022**, *64*, 1352–1363. [CrossRef]
49. Wei, X.; Zhou, H.; Xie, D.; Li, J.; Yang, M.; Chang, T.; Wang, D.; Hu, L.; Xie, G.; Wang, J.; et al. Genome-Wide Association Study in Rice Revealed a Novel Gene in Determining Plant Height and Stem Development, by Encoding a WRKY Transcription Factor. *Int. J. Mol. Sci.* **2021**, *22*, 8192. [CrossRef]
50. Liu, X.; Zhou, C.; Zhao, Y.; Zhou, S.; Wang, W.; Zhou, D.X. The rice enhancer of zeste [E(z)] genes SDG711 and SDG718 are respectively involved in long day and short day signaling to mediate the accurate photoperiod control of flowering time. *Front. Plant Sci.* **2014**, *5*, 591. [CrossRef] [PubMed]
51. Gao, H.; Jin, M.; Zheng, X.M.; Chen, J.; Yuan, D.; Xin, Y.; Wang, M.; Huang, D.; Zhang, Z.; Zhou, K.; et al. Days to heading 7, a major quantitative locus determining photoperiod sensitivity and regional adaptation in rice. *Proc. Natl. Acad. Sci. USA* **2014**, *111*, 16337–16342. [CrossRef] [PubMed]
52. Cho, L.H.; Yoon, J.; Wai, A.H.; An, G. Histone Deacetylase 701 (HDT701) Induces Flowering in Rice by Modulating Expression of OsIDS1. *Mol. Cells* **2018**, *41*, 665–675. [CrossRef] [PubMed]

53. Wright, S. Variability within and among natural populations. In *Evolution and the Genetics of Populations*; University of Chicago Press: Chicago, IL, USA, 1978; Volume 4.
54. Yan, W.H.; Wang, P.; Chen, H.X.; Zhou, H.J.; Li, Q.P.; Wang, C.R.; Ding, Z.H.; Zhang, Y.S.; Yu, S.B.; Xing, Y.Z.; et al. A major QTL, Ghd8, plays pleiotropic roles in regulating grain productivity, plant height, and heading date in rice. *Mol. Plant* **2011**, *4*, 319–330. [CrossRef] [PubMed]
55. Wei, X.; Xu, J.; Guo, H.; Jiang, L.; Chen, S.; Yu, C.; Zhou, Z.; Hu, P.; Zhai, H.; Wan, J. DTH8 suppresses flowering in rice, influencing plant height and yield potential simultaneously. *Plant Physiol.* **2010**, *153*, 1747–1758. [CrossRef]
56. Uno, Y.; Furihata, T.; Abe, H.; Yoshida, R.; Shinozaki, K.; Yamaguchi-Shinozaki, K. Arabidopsis basic leucine zipper transcription factors involved in an abscisic acid-dependent signal transduction pathway under drought and high-salinity conditions. *Proc. Natl. Acad. Sci. USA* **2000**, *97*, 11632–11637. [CrossRef]
57. Wang, Y.; Lu, Y.; Guo, Z.; Ding, Y.; Ding, C. RICE CENTRORADIALIS 1, a TFL1-like Gene, Responses to Drought Stress and Regulates Rice Flowering Transition. *Rice* **2020**, *13*, 70. [CrossRef]
58. Izawa, T. Adaptation of flowering-time by natural and artificial selection in Arabidopsis and rice. *J. Exp. Bot.* **2007**, *58*, 3091–3097. [CrossRef]
59. Luan, W.; Chen, H.; Fu, Y.; Si, H.; Peng, W.; Song, S.; Liu, W.; Hu, G.; Sun, Z.; Xie, D.; et al. The effect of the crosstalk between photoperiod and temperature on the heading-date in rice. *PLoS ONE* **2009**, *4*, e5891. [CrossRef]
60. Zong, W.; Ren, D.; Huang, M.; Sun, K.; Feng, J.; Zhao, J.; Xiao, D.; Xie, W.; Liu, S.; Zhang, H.; et al. Strong photoperiod sensitivity is controlled by cooperation and competition among Hd1, Ghd7 and DTH8 in rice heading. *New Phytol.* **2021**, *229*, 1635–1649. [CrossRef]

 *agronomy*

Article

Responses of the Lodging Resistance of *Indica* Rice Cultivars to Temperature and Solar Radiation under Field Conditions

Xiaoyun Luo [1,2], Zefang Wu [3], Lu Fu [1,2], Zhiwu Dan [1,2], Weixiong Long [1,2], Zhengqing Yuan [1,2], Ting Liang [1,2], Renshan Zhu [1,2], Zhongli Hu [1,2] and Xianting Wu [1,2,3,*]

[1] State Key Laboratory of Hybrid Rice, Wuhan University, Wuhan 430072, China
[2] College of Life Sciences, Wuhan University, Wuhan 430072, China
[3] Crop Research Institute, Sichuan Academy of Agricultural Science, Chengdu 610000, China
* Correspondence: xiantwu@whu.edu.cn

Abstract: Much attention has shifted to the effects of temperature and solar radiation on rice production and grain quality due to global climate change. Meanwhile, lodging is a major cause of rice yield and quality losses. However, responses of the lodging resistance of rice to temperature and solar radiation are still unclear. To decipher the mechanisms through which the lodging resistance might be affected by temperature and solar radiation, 32 rice cultivars with different lodging resistance were grown at two eco-sites on three sowing dates over a period of three years. Based on the field observation, 12 *indica* rice cultivars which did not lodge were selected for analysis. Significant differences were found in the lodging resistance of the *indica* rice cultivars at different temperature and solar radiation treatments. The results showed that temperature was the main factor that affected the lodging resistance of *indica* rice cultivars under the conditions of this study. With the increased average daily temperature, the lodging resistance decreased rapidly, primarily due to the significant reduction in physical strength of the culm, which was attributed to the longer and thinner basal second internode. Among the 12 *indica* rice cultivars, the lodging-moderate cultivar Chuanxiang 29B was most sensitive to temperature, and the lodging-resistant cultivar Jiangan was least responsive to temperature. These results suggested that rice breeders could set the shorter and thicker basal internode as the main selection criteria to cultivate lodging-resistant *indica* cultivars to ensure a high yield at a higher ambient temperature.

Keywords: temperature; solar radiation; sowing date; growth duration; lodging resistance; lodging index; lodging-related traits; *indica* rice

Citation: Luo, X.; Wu, Z.; Fu, L.; Dan, Z.; Long, W.; Yuan, Z.; Liang, T.; Zhu, R.; Hu, Z.; Wu, X. Responses of the Lodging Resistance of *Indica* Rice Cultivars to Temperature and Solar Radiation under Field Conditions. *Agronomy* **2022**, *12*, 2603. https://doi.org/10.3390/agronomy12112603

Academic Editor: Roberto Barbato

Received: 30 August 2022
Accepted: 20 October 2022
Published: 23 October 2022

Publisher's Note: MDPI stays neutral with regard to jurisdictional claims in published maps and institutional affiliations.

1. Introduction

Lodging severely reduces the grain yield and quality of rice [1]. Furthermore, it increases production costs by adversely affecting the harvest manipulations and heightening the grain drying demand [2,3]. According to a study by Nakajima et al. [4], rice lodging could also aggravate mycotoxin pollution that threatens animal and human health. Since the initiation of the "Green revolution" in the 1960s, semi-dwarf cultivars of rice and wheat have been developed, which have enhanced the lodging resistance significantly and increased global grain production [5–7]. However, with the large-scale cultivation of high-yielding cultivars, extensive use of fertilizers, and simplified planting techniques, such as direct-seeding, the potential risk of lodging has increased in recent years [8–10]. Thus, lodging-resistant cultivars have been developed as a genetic improvement strategy to increase the yield of rice, wheat, and other crops [11–14].

Lodging, which results from a loss of balance in plant bodies, refers to the lasting vertical stem displacement of plants [15]. In the case of rice, there are three types of lodging: culm bending, culm breaking and root lodging [3,16]. Culm breaking is generally seen at the lower internodes (including the third and fourth internodes from the plant top), which happens when the bending moment of the upper plant part is excessive [2,17]. The manner in which rice resists against culm breaking is usually assessed by the lodging index (LI) [18–21]. A decrease in the LI indicates a stronger lodging resistance capability. Several studies have investigated the correlation of LI with lodging-related traits [22–24]. However, as the LI is a ratio of the bending moment of the whole plant (BM) to the bending moment at breaking (M), it is inadequate for evaluating the lodging resistance capability under certain special conditions. For example, when the multiple differences in BM values between two cultivars are similar to the multiple differences in M values between the two cultivars, the calculated LI values of the two cultivars would exhibit no significant difference, which could be contradictory to the actual lodging resistance of the rice cultivars. As a result, an optimized parameter $\triangle$BM (equal to the value of 2M minus the value of BM), which was defined as the external force that the basal second internode could withstand, was proposed to be used along with the LI for a further accurate evaluation of the lodging resistance [25]. An increase in the $\triangle$BM indicates a stronger lodging resistance capability. Lodging generally happens at the stage of grain filling [26,27]. In the research of Ichii and Hada [28], the stem-breaking strength decreased to the minimum value and the LI increased to the maximum at 30 days after heading, which indicated that the grain-filling stage is the period when lodging often occurs. Lodging is associated with several biotic and abiotic factors: the height and weight of the plant, the length of the panicle, the plumpness of the leaf sheath, as well as the length, diameter and thickness of basal internodes influence the lodging resistance of rice [29–34]. Regarding the morphological factors, the external diameter and thickness of the cross-section from basal internodes strongly influence the breaking strength of the stem [21,35]. Additionally, the stem contents of soluble sugars, K, Si, cellulose, starch and lignin affect the basal stem-breaking strength pronouncedly [36–40]. However, some researchers have suggested that the basal stem-breaking strength is decided by structural carbohydrates (lignin and cellulose) rather than non-structural carbohydrates (soluble sugars and starch) [20,41]. Growth conditions, such as the application of different fertilizers, planting density, direct seeding methods, and sheath blight attacks, strongly affect the lodging resistance of rice plants [10,19,42–45]. Lodging is also correlated with varying environmental parameters, such as rain, wind, CO_2, deep water, and resource complementarities [46–50].

With the increase in depletion of the stratospheric ozone, atmospheric levels of greenhouse gases, land-use alterations and aerosol outputs, an increase in global temperatures (global warming) and a decrease in solar radiation in Asia have been recorded in recent decades [51–54]. In the last century, the average global surface temperature recorded an elevation by 0.5 °C, and its estimated range of elevation is 0.3–6.4 °C by the end of this century [55]. An average annual reduction of 0.51 ± 0.05 W m^{-2} in solar radiation in Asia has been reported [56,57]. Many studies have shown a significant influence of the increase in global temperatures on the yield and quality of rice grains [58,59]. Additionally, the positive role and significance of solar radiation in rice grain output have also been shown [60,61]. However, few studies have considered the influences imposed on rice lodging resistance by the solar radiation and temperature variations. One study reported that an increase in the soil temperature increased the lodging risk of rice plants [18]. The low solar radiation reduced the physical strength of the stem and, thus, increased lodging susceptibility in rice [62].

More than half of the global population consumes rice as a staple food. Since lodging, high temperature, and low solar radiation have detrimental effects on rice production, understanding the effects of temperature and solar radiation on lodging is important for growing rice that is adapted to the changing global climate. In the present work, we chose two eco-sites to carry out 3-year field experimentations on three sowing dates per year, to grow rice at different temperatures and under different solar radiation treatments. A total of 32 rice cultivars with different lodging resistance capabilities were evaluated under different combinations of temperature and solar radiation. Among them, 12 *indica* rice cultivars, which did not lodge in all the sowing dates based on the field observation, were selected for analysis. The objectives of the present study were to: (1) investigate the responses of the lodging resistance of *indica* rice cultivars to different temperature and solar radiation treatments; (2) evaluate the most and least affected cultivars among the 12 *indica* rice cultivars under different temperature and solar radiation treatments; (3) explore the relationship of the morphological, mechanical, and biochemical characteristics associated with lodging resistance with temperature and solar radiation. To cope with global climate change, a greater understanding of the climatic impact on lodging in *indica* rice will provide guidelines for rice breeders to adopt appropriate strategies for developing lodging-resistant *indica* rice cultivars in the future.

2. Materials and Methods

2.1. Experimental Materials

Thirty-two rice cultivars in total, including 18 Chinese accessions, 11 cultivars from a mini-core subset of the United States Department of Agriculture (USDA) rice gene bank [63], as well as Kasalath from India, Lemont from the USA and IR58025B from the IRRI, were selected as the experimental materials. These 32 rice cultivars were categorized into three groups (Groups 1, 2 and 3) with different lodging-resistance capabilities based on the principal component analysis and hierarchical clustering analysis of the lodging index in our previous study [25]. Group 1 (the lodging-susceptive group) comprised 3 *indica* cultivars and 4 *japonica* cultivars. There were 18 *indica* cultivars and four *japonica* cultivars in Group 2 (the lodging-moderate group). The remaining three rice cultivars in Group 3 (the lodging-resistant group) all belong to *indica*. Table S1 lists the detailed grouping information.

2.2. Experimental Design

For phenotypic information acquisition of the tested cultivars at varying temperatures and solar radiations, we chose two eco-sites to carry out 3-year field experiments on three sowing dates (SDs) per year, which totaled three experiments. The location of the first experiment was Xindu in China's Sichuan province (30°49′51.51″ N, 104°06′3.44″ E; 547.7 m altitude), and the experiment was conducted during the rice-growing season from April to September in 2015. The location of the other two experiments was Ezhou in China's Hubei province (30°22′20.75″ N, 114°45′7.78″ E; 23.6 m altitude), and the experiments were conducted during the rice-growing seasons from May to October in 2017 and 2018, respectively. Every experiment was arranged in a split plot design with sowing dates (SDs) as main plots and cultivars as subplots. The 32 cultivars were all sown on April 11 (SD1), April 20 (SD2) and April 28 (SD3) in 2015 at the Xindu site. Following growth to about the fourth leaf stage, we transplanted seedlings from each SD to paddies in an experimental block roughly sizing 16 m × 6 m (96 m^2). Three blocks were set up for the three SDs. Each block was randomly arranged with 32 plots (each 2 m × 0.6 m in size; 1.2 m^2) for the 32 cultivars. Each cultivar was planted in triplicate rows (10 plants per row) at each experimental plot, where the hill spacing was 20 cm × 20 cm. Isolation of consecutive plots was accomplished at one-row spacing of 20 cm so that the growth impact on adjacent plants could be minimized and the plant cultivar could be clarified. In the case of the Ezhou site, the 32 cultivars were all sown on May 8 (SD4), May 23 (SD5) and June 7 (SD6) in 2017 for the second experiment, and on May 10 (SD7), May 25 (SD8) and June 9 (SD9) in 2018 for the third experiment. The seedling transplantation and plot designs for each SD were identical to those for Xindu.

In each experiment, 375 kg ha^{-1} of basal fertilizer in total, which was the compound fertilizer containing each 15% *w/w* N, P and K, was applied to each block one day before the seedling transplantation. As an N source, urea was split-applied during the tillering stage at 150 kg ha^{-1} and during the panicle initiation stage at 75 kg ha^{-1}. About a 5-cm water depth was maintained in the experimental field post-transplant and remained flooded until 7 days before maturity. Intensive control of diseases, insects and weeds was implemented, in order to avoid biomass or yield loss.

2.3. Meteorological Data Collection

The sources of the sunshine hours and temperature data, including the daily maximum, average and minimum temperatures, were the China Meteorological Data Service Centre [64] and the relevant local meteorological stations near the experimental fields (Figure 1). Due to the lack of equipment for measuring solar radiation, the solar radiation data could not be directly recorded at the meteorological stations. Therefore, an Angstrom empirical model was employed to simulate the 2015 solar radiation data for Xindu, and the 2017 and 2018 solar radiation data for Ezhou based on the sunshine hours [65,66]:

$$R_G/R_A = a + b \times n/N \tag{1}$$

where R_G and R_A, respectively, denote the global and extraterrestrial solar radiations (both MJ m^{-2} day^{-1}), whereas n and N, respectively, represent the actual and potential daily sunshine hours. Following the method in a study by Chen et al. [67], "a" and "b", the empirical factors, were found to be 0.15 and 0.55 at Xindu, and 0.13 and 0.52 at Ezhou, respectively. The Nash–Sutcliffe equation (NSE) values were 0.81 and 0.84 for Xindu and Ezhou, respectively, which indicated that the model ran well [68].

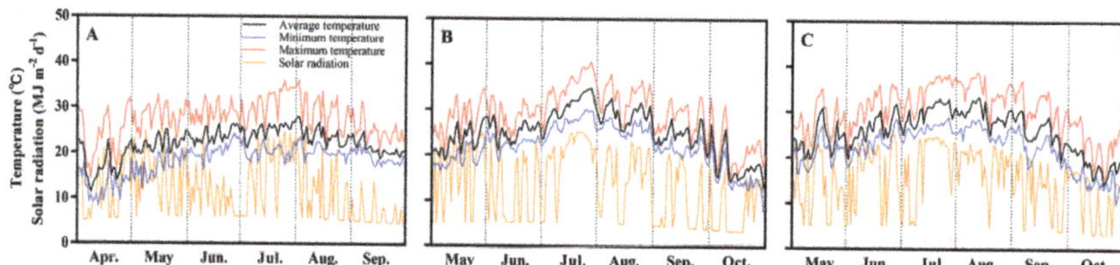

Figure 1. The daily average, minimum, and maximum temperatures and solar radiation during the whole rice growth season in 2015 at Xindu (**A**), and in 2017 (**B**) and 2018 (**C**) at Ezhou.

The effective accumulated temperature (EAT) in the determined growth duration was calculated as:

$$\text{EAT} \ (^{\circ}\text{C}) = \sum (T - T_0) \times \text{Growth duration (d)} \tag{2}$$

where T and T_0 (10 °C for rice) are the daily average temperature and the biological zero temperature, respectively [58,69].

The cumulative solar radiation (CSR) in the determined growth duration was calculated as:

$$\text{CSR} \ (\text{MJ m}^{-2}) = \sum R \times \text{Growth duration (d)} \tag{3}$$

where R (MJ m^{-2} d^{-1}) is the daily solar radiation, which is calculated as the R_G from the formula (1).

2.4. Plant Sampling and Measurements

The growth period, which spanned from the sowing to the maturation stages, varied from 90 to 146 days depending on the different cultivars and environmental factors. According to the cultivar growth period in each plot, we recorded the full-heading date, which was defined as the date on which 80% emergence of all panicles occurred from the flag leaf sheath. At both experimental sites, we assessed the lodging-related traits of the plants in each plot 25 days following the full-heading date, i.e., the grain filling stage. The days of the determined growth duration from the sowing date to the measuring date were also recorded for each cultivar plot (Table 1). For the phenotypic determination, eight plants were picked from the middle of each plot to avoid the marginal effects. These harvested plants, which had all tillers and intact roots, were transferred to the laboratory in plastic buckets filled with water for subsequent analyses. For lodging-related traits determination, the representative samples of plant main tillers were collected. We examined only the traits associated with the second elongated internode (starting from the main tiller root) in view of the common occurrence of culm-breaking-lodging type at the lower internodes [2,70]. Measurement of the culm length (CL) was accomplished between the lower node of the basal second internode and the panicle tip. Measurement of the stem length (SL) was accomplished between the lower nodes of the basal second and third internodes. Panicle length (PL) measurement was accomplished from the bottom node to the tip of a panicle. Determination of the fresh weight (FW) was accomplished for the zone between the lower node of the basal second internode and the panicle tip. A YYD-1 plant lodging tester (TOP Instrument, Zhejiang China) was utilized for determining the breaking resistance (BR) at the middle point of the basal second internode with a leaf sheath as per a priorly reported procedure [29]. Each sampled basal second internode with a leaf sheath was placed on the groove of supporting pillars that were 5 cm apart. The plant lodging tester, which was arranged perpendicularly to the middle internode, was then progressively loaded. BR measurement was accomplished when the internode was pushed to break, and the tester readout was recorded as the BR value (kg). Measurements of the culm diameter (CD), culm wall thickness (CT), as well as the external and internal diameters of the minor and major axes in an oval cross-section at the basal second internode center were accomplished using a digital vernier caliper. The physical parameters were calculated following the studies of Ookawa and Ishihara [71] and Ookawa et al. [72], and our previous study [25] as follows:

$$\text{Bending moment of the whole plant (BM, g cm)} = \text{CL (cm)} \times \text{FW (g)} \tag{4}$$

where CL is the culm length from the lower node of the basal second internode to the panicle tip, FW is the fresh weight from the lower node of the basal second internode to the panicle tip.

$$\text{Bending moment at breaking (M, g cm)} = 1/4 \times \text{BR (kg)} \times \text{Spacing between supporting pillars (cm)} \times 10^3 \tag{5}$$

where BR is the breaking resistance of the basal second internode with leaf sheath; spacing between supporting pillars is 5 cm.

$$\text{Lodging index (LI, \%)} = \text{BM}/\text{M} \times 100\% \tag{6}$$

where BM and M are calculated from formulas (4) and (5), respectively.

$$\text{The external force that the basal second internode could withstand } (\triangle\text{BM, g cm}) = 2\text{M} - \text{BM} \tag{7}$$

where BM and M are calculated from formulas (4) and (5), respectively.

$$\text{Section modulus (SM, mm}^3) = \pi/32 \times (a_1{}^3 b_1 - a_2{}^3 b_2)/a_1 \tag{8}$$

$$\text{Culm diameter (CD, mm)} = (a_1 + b_1)/2 \tag{9}$$

$$\text{Culm wall thickness (CT, mm)} = (a_1 - a_2 + b_1 - b_2)/4 \tag{10}$$

where a_1 and b_1 respectively denote the outer diameters of the minor and major axes in an oval cross-section, whereas a_2 and b_2 respectively represent the inner diameters of the minor and major axes in an oval cross-section.

$$\text{Bending stress (BS, g mm}^{-2}) = \text{M}/\text{SM} \times 10 \tag{11}$$

where M and SM are calculated from formulas (5) and (8), respectively.

Following the determination of all lodging-related parameters, the basal second internode with the leaf sheath was subjected to 105 °C oven-drying for 30 min, then dried at 70 °C until the weight was constant. The dried basal second internodes were then ground to a fine powder for determining structural carbohydrates. The cellulose content (CC) was measured using a commercial cellulose content assay kit (Boxbio Science, Beijing, China), which was modified following the method of Updegraff [73]. Since the acetyl bromide method for the determination of lignin appears to have earned the most widespread acceptance [74], we used a commercial lignin content assay kit (Boxbio Science, Beijing, China), which was modified following the method of Johnson et al. [75], to measure the lignin content (LC) of the samples. The CD, CT, LC and CC were only measured in 2018 at Ezhou.

Table 1. The determined growth durations (days) of twelve *indica* rice cultivars on all the sowing dates across two locations and three years.

Cultivars	Growth Duration (d)								
	Xindu/2015			Ezhou/2017			Ezhou/2018		
	SD1 [a]	SD2	SD3	SD4	SD5	SD6	SD7	SD8	SD9
Chuan 106B	112	108	106	104	101	102	102	100	101
345B	114	113	109	106	105	104	106	105	102
Huanghuazhan	126	123	119	109	107	105	110	105	105
Jinlongsimiao	126	121	117	115	112	107	111	108	102
Chuanxiang 29B	124	126	119	112	109	107	112	107	106
Chenghui 3203	136	129	122	116	114	105	113	107	108
Guichao 2	126	121	117	111	108	106	111	106	103
II-32B	128	127	121	116	121	111	117	116	113
Teqing	128	125	119	116	114	106	114	107	107
R379	144	139	134	125	122	116	125	121	115
9311	129	124	121	112	111	106	112	110	108
Jiangan	147	142	136	117	115	107	118	110	106

[a] SD1, 11 April 2015; SD2, 20 April 2015; SD3, 28 April 2015; SD4, 8 May 2017; SD5, 23 May 2017; SD6, 7 June 2017; SD7, 10 May 2018; SD8, 25 May 2018; SD9, 9 June 2018. All abbreviations imply the same below as well.

2.5. Data Analysis

According to the field observation, 12 *indica* rice cultivars, including the nine cultivars (Chuan 106B, 345B, Huanghuazhan, Jinlongsimiao, Chuanxiang 29B, Chenghui 3203, Guichao 2, II-32B and Teqing) in the lodging-moderate group and the three cultivars (R379, 9311 and Jiangan) in the lodging-resistant group, which did not lodge on all the sowing dates, were selected for the following data analysis. The measured data were analyzed via the Statistical Product and Service Solutions software for Windows (SPSS, Ver. 20.0; IBM, NY, USA). For each rice cultivar, the mean of eight main tillers was determined as the lodging-related parameter for that cultivar plot. The average daily temperature (T_{mean}) and solar radiation (R_{mean}), effective accumulated temperature (EAT) and cumulative solar radiation (CSR) for each SD were calculated based on the mean values of all the 12 *indica* rice cultivars in the determined growth durations. A Tukey test was employed to accomplish the significance analysis of the T_{mean}, R_{mean}, EAT, CSR, LI and $\triangle$BM for the different sowing dates in each year, where $p = 0.05$ was set as the statistical significance level. To determine the relationship of the temperature and solar radiation parameters with the lodging-related parameters, we minimized the inter-cultivar differences in traits using the standardized data according to the method described in a study by Fan and Liu [76]. The standardized data, which were referred to as "relative data", were calculated according to the formula below:

The relative lodging index = the lodging index of a given cultivar on one sowing date/average lodging index of this cultivar on all the sowing dates.

The other relative values of lodging-related parameters were calculated using the formula for the relative lodging index described above. The stepwise regression analysis and path analysis were performed to determine the relationships of temperature and solar radiation parameters to the lodging resistance. The correlation analysis was conducted using Pearson's r coefficient, and the correlations were regarded as statistically significant when $p < 0.05$. The correlation coefficient was decomposed into direct effects and indirect effects by path analysis, with the result that the relative importance of the temperature and solar radiation for the lodging resistance were revealed. The values of direct and indirect effects were calculated according to the study of Luo et al. [77]. The GraphPad Prism software for Windows (Ver. 5.0; San Diego, CA, USA) was utilized for depicting the scatter diagrams and histograms.

3. Results

3.1. Effects of Sowing Date on Determined Growth Durations and Temperature and Solar Radiation Conditions

The determined growth durations of all the 12 *indica* rice cultivars were shortened as the sowing dates were delayed in each year (Table 1). Compared with the first sowing date, the number of days of the determined growth durations in the last sowing date were reduced by 5−14 d, 2−11 d, and 1−12 d in 2015, 2017, and 2018, respectively. The average determined growth durations of the three sowing dates at Ezhou were reduced by 6−30 d compared with those at Xindu.

The average daily temperature (T_{mean}) of the experiment during the determined growth durations increased significantly with the delay of the sowing dates in each year (Figure 2A). T_{mean} on SD1 was 0.4 and 0.8 °C lower than that on SD2 and SD3, respectively. T_{mean} in SD4 was 0.2 and 0.3 °C lower than that on SD5 and SD6, respectively. Additionally, T_{mean} on SD7 was 0.5 and 0.8 °C lower than that on SD8 and SD9, respectively. Furthermore, the T_{mean} showed significant differences between locations and years (Table S2). The T_{mean} from SD1 to SD3 in 2015 at Xindu was significantly lower than that from SD4 to SD6 in 2017 and from SD7 to SD9 in 2018 at Ezhou by 4.3 and 5.1 °C, respectively. However, the effective accumulated temperature (EAT) showed no significant variations among the three sowing dates in each year. Similar to the T_{mean}, the EAT at Xindu was significantly lower than that at Ezhou (Figure 2B, Table S2).

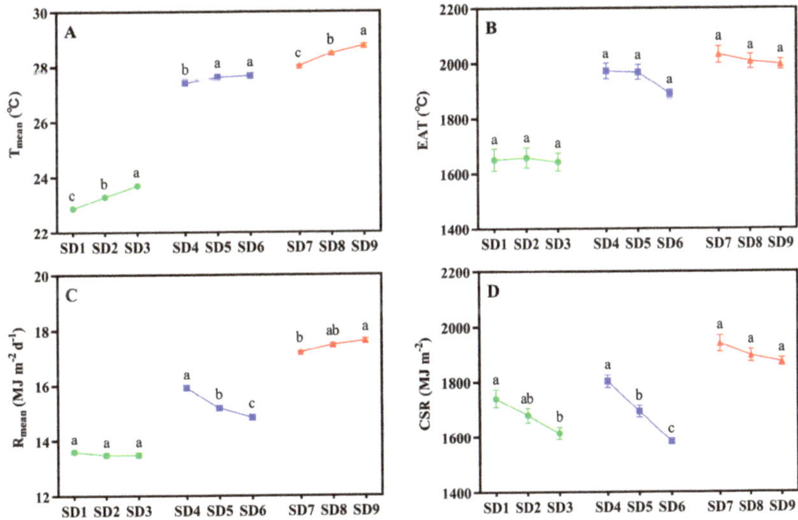

Figure 2. Differences in the average daily temperature (T_{mean}) (**A**) and solar radiation (R_{mean}) (**B**), effective accumulated temperature (EAT) (**C**) and cumulative solar radiation (CSR) (**D**) of the twelve *indica* rice cultivars on the three sowing dates of each year during the determined growth durations. SD1, 11 April 2015; SD2, 20 April 2015; SD3, 28 April 2015; SD4, 8 May 2017; SD5, 23 May 2017; SD6, 7 June 2017; SD7, 10 May 2018; SD8, 25 May 2018; SD9, 9 June 2018. All abbreviations imply the same below as well. Vertical bars indicate standard errors (±), n = 12 for each SD. Different lowercase letters on the bars of the three SDs in each year indicate significant differences determined by the Tukey test at 5% probability level.

The average solar radiation (R_{mean}) of the experiment during the determined growth durations showed an inconsistent tendency with the delayed sowing dates in each year

(Figure 2C). From SD1 to SD3, the R_{mean} was relatively constant. However, the R_{mean} on SD4 was 0.7 and 1.1 MJ m^{-2} day^{-1} higher than that on SD5 and SD6, respectively. Compared to the declining trend from SD4 to SD6, the R_{mean} displayed the opposite trend from SD7 to SD9, where the R_{mean} on SD7 was lower than that on SD8 and SD9 by 0.3 and 0.4 MJ m^{-2} day^{-1}. In addition, the R_{mean} was also significantly affected by locations and years (Table S2). The R_{mean} from SD1 to SD3 in 2015 at Xindu was significantly lower than that from SD4 to SD6 in 2017 and from SD7 to SD9 in 2018 at Ezhou by 1.8 and 3.9 MJ m^{-2} day^{-1}, respectively. As a result of delayed sowing dates, the cumulative solar radiation (CSR) decreased significantly from SD1 to SD3 and from SD4 to SD6 (Figure 2D). In addition, the CSR displayed an insignificant declining trend from SD7 to SD9. On the other hand, the CSR in 2015 at Xindu and in 2017 at Ezhou were significantly lower than those in 2018 at Ezhou (Figure 2D, Table S2).

Thus, a total of nine temperature and solar radiation treatments with significant differences were established for the twelve *indica* rice cultivars by setting nine sowing dates across two eco-sites and three years.

3.2. Effects of Sowing Date on LI and △BM

The LI and △BM, which were used to evaluate the lodging resistance of rice, were significantly affected by the sowing dates (Figure 3). With the delayed sowing dates, the LI increased significantly in each year (Figure 3A). The LI on SD3 was 22.65% higher than SD1. The LI on SD6 was 19.82% higher than SD4. Additionally, the LI on SD9 was 17.58% higher than SD7. Although the tendency of LI was not always significant in some cultivars, the overall increasing tendencies of LI of the 12 *indica* cultivars were identical (Table 2). In addition, the LI was also significantly affected by locations and years (Table S2). The average LI of the three SDs (SD1, SD2, and SD3) in 2015 at Xindu was significantly lower than that of the three SDs (SD4, SD5, and SD6) in 2017 and the three SDs (SD7, SD8, and SD9) in 2018 at Ezhou by 41.03% and 36.97%, respectively. Contrary to the LI, the △BM decreased significantly with the delay in the sowing dates in each year (Figure 3B). Compared to the first sowing date, the △BM of the last sowing date was lower by 35.52%, 57.62% and 47.93% in 2015, 2017 and 2018, respectively. The decreasing trends of △BM of all the 12 *indica* cultivars were significant (Table 3). Furthermore, the average △BM of the three SDs in 2015 at Xindu was significantly higher than that of the three SDs in 2017 and 2018 at Ezhou by 255.88% and 179.51%, respectively. However, there was no significant difference in the △BM between 2017 and 2018 at Ezhou (Table S2). These results indicated that the lodging resistance of 12 *indica* rice cultivars weakened with the delay in the sowing dates in each year and it was greater at Xindu than that at Ezhou.

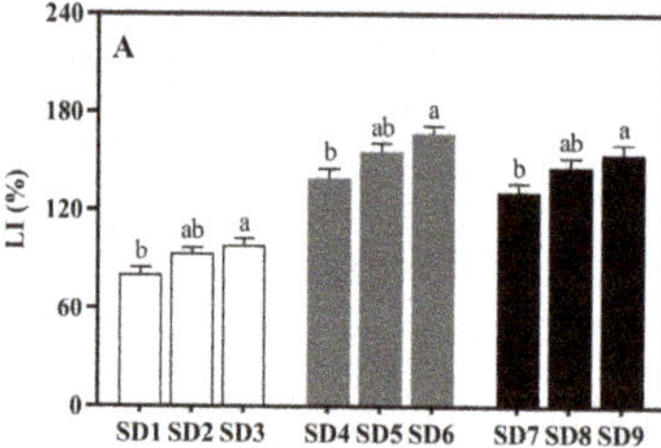
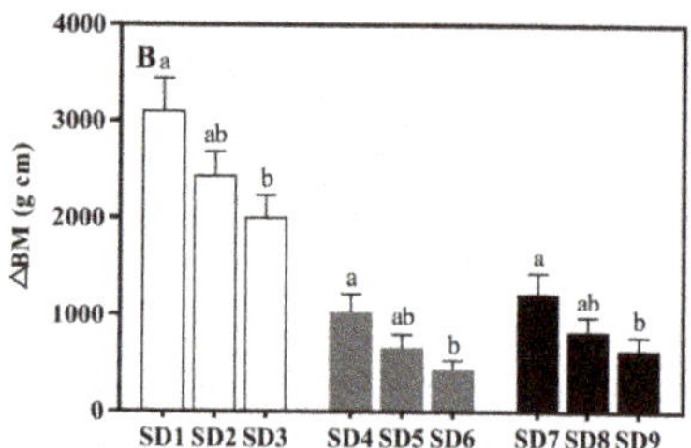

Figure 3. Comparison of the LI (**A**) and △BM (**B**) for the three sowing dates in each year. LI, lodging index; △BM, the external force that the basal second internode could withstand. All abbreviations imply the same below as well. Vertical bars indicate standard errors ($\pm$), n = 12 for each SD. Different lowercase letters on the bars indicate significant differences determined by the Tukey test at 5% probability level.

Table 2. The lodging index of twelve *indica* rice cultivars on all the sowing dates across two locations and three years.

Cultivars	Lodging Index (%)								
	Xindu/2015			Ezhou/2017			Ezhou/2018		
	SD1	SD2	SD3	SD4	SD5	SD6	SD7	SD8	SD9
Chuan 106B	47.11 b [a]	69.62 a	78.72 a	107.73 c	143.22 b	164.70 a	117.37 b	139.24 a	147.66 a
345B	63.46 b	88.30 a	89.25 a	123.65 b	138.31 ab	147.49 a	117.79 b	122.94 b	142.64 a
Huanghuazhan	70.98 b	69.87 b	84.58 a	136.87 b	165.57 a	181.87 a	116.95 b	137.56 a	150.58 a
Jinlongsimiao	78.09 b	86.15 ab	93.18 a	145.71 b	161.86 ab	178.79 a	125.62 b	152.64 a	154.23 a
Chuanxiang 29B	80.84 b	99.29 a	94.52 a	163.60 a	169.84 a	178.62 a	155.36 b	165.75 ab	179.44 a
Chenghui 3203	83.25 a	92.10 a	91.25 a	151.27 b	159.50 ab	176.02 a	145.68 a	153.73 a	152.95 a
Guichao 2	83.72 a	86.04 a	93.91 a	131.29 b	161.73 a	163.69 a	127.07 b	139.46 a	148.41 a
II-32B	89.32 b	109.54 a	114.86 a	162.52 a	166.74 a	184.31 a	155.90 b	178.68 a	186.06 a
Teqing	99.70 b	110.16 b	127.81 a	155.11 a	172.41 a	172.32 a	147.48 c	162.21 b	177.00 a
R379	86.56 a	100.41 a	98.20 a	127.60 b	142.23 a	152.40 a	118.40 b	131.65 a	127.62 ab
9311	91.19 b	106.94 a	117.20 a	160.74 a	173.83 a	170.69 a	149.48 b	162.77 a	167.92 a
Jiangan	94.39 b	103.09 a	104.54 a	110.10 b	118.95 b	137.43 a	103.25 b	115.53 a	123.70 a

[a] Within a row for each year, means followed by the same letters are not significantly different determined by the Tukey test at 5% probability level.

Table 3. The $\triangle$BM of twelve *indica* rice cultivars on all the sowing dates across two locations and three years.

Cultivars	$\triangle$BM (g cm)									CV [a] (%)
	Xindu/2015			Ezhou/2017			Ezhou/2018			
	SD1	SD2	SD3	SD4	SD5	SD6	SD7	SD8	SD9	
Chuan 106B	2644.12 a [b]	2039.94 b	1346.41 c	1150.50 a	671.60 b	349.58 c	1177.42 a	758.29 b	517.23 b	63.19
345B	2501.73 a	1790.85 b	1642.24 b	922.84 a	617.56 b	477.06 b	1078.02 a	926.45 a	642.13 b	56.76
Huanghuazhan	3149.41 a	2629.95 ab	2036.62 b	736.20 a	341.49 b	146.86 c	1082.53 a	660.49 b	473.25 b	86.74
Jinlongsimiao	3123.09 a	2740.98 a	2137.34 b	811.33 a	550.77 ab	292.45 b	1176.19 a	733.90 b	635.89 b	76.76
Chuanxiang 29B	2600.97 a	2163.80 b	1957.99 b	540.39 a	416.68 ab	264.07 b	684.26 a	481.05 ab	250.41 b	88.95
Chenghui 3203	3174.04 a	2443.57 b	2444.77 b	850.97 a	591.92 ab	300.71 b	1003.42 a	696.71 b	695.78 b	76.57
Guichao 2	2213.49 a	2108.17 a	1613.65 b	897.83 a	420.38 b	353.33 b	1046.41 a	725.54 b	551.51 b	64.28
II-32B	1793.73 a	1436.09 ab	1176.57 b	480.55 a	354.43 ab	159.34 b	559.40 a	302.48 ab	133.49 b	85.05
Teqing	1891.21 a	1511.49 b	1039.69 c	586.78 a	329.64 b	312.78 b	703.07 a	451.53 b	252.74 c	73.69
R379	5442.30 a	3728.34 b	3329.22 b	1836.23 a	1144.77 b	847.96 c	2181.02 a	1483.45 b	1454.91 b	62.88
9311	3678.55 a	2560.75 b	1839.95 c	727.80 a	407.82 b	461.61 ab	1043.13 a	663.80 b	488.15 b	86.82
Jiangan	5089.65 a	4192.39 ab	3563.15 b	2838.87 a	2033.76 b	1280.59 c	3062.08 a	2090.83 b	1609.43 c	43.87

[a] CV, coefficient of variation of the $\triangle$BM across the nine sowing dates. [b] Within a row for each year, means followed by the same letters are not significantly different determined by the Tukey test at 5% probability level.

3.3. Relationship of Temperature and Solar Radiation Parameters with Relative LI and $\triangle$BM

The relative LI increased significantly with the T_{mean} in each year (Figure 4A). Contrastingly, the relative $\triangle$BM decreased significantly with the T_{mean} in each year (Figure 4E). No significant correlations between the EAT and the relative LI or $\triangle$BM were observed in each year (Figure 4B,F). Significant linear correlations between the R_{mean} and the relative LI and $\triangle$BM were found in each year except in 2015 (Figure 4C,G). However, the linear correlations were reversed in 2017 and 2018. The relative LI showed a significantly negative correlation with the CSR in 2015 and 2017 (Figure 4D). The relative $\triangle$BM displayed the opposite result, as was expected (Figure 4H). However, both the relative LI and $\triangle$BM showed no significant correlation with the CSR in 2018.

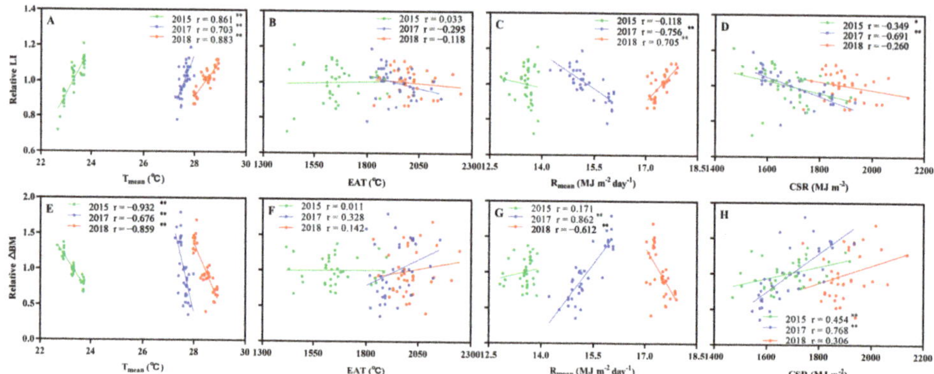

Figure 4. Correlations between the relative LI of the twelve *indica* rice cultivars and average daily temperature (T_{mean}) (**A**) and solar radiation (R_{mean}) (**B**), effective accumulated temperature (EAT) (**C**) and cumulative solar radiation (CSR) (**D**) on the three sowing dates of each year. Correlations between the relative $\triangle$BM of the twelve *indica* rice cultivars and T_{mean} (**E**), R_{mean} (**F**), EAT (**G**), and CSR (**H**) on the three sowing dates of each year. The green, blue, and red points represent the cultivars measured in 2015 at Xindu, and in 2017 and 2018 at Ezhou, respectively. $n = 36$ in each year. * and **, significant differences at $p < 0.05$ and $p < 0.01$, respectively.

When the nine temperature and solar radiation treatments were combined, significant correlations were also found between the temperature and solar radiation parameters and the relative LI and $\triangle$BM (Figure 5). There were significant positive linear correlations between the relative LI and T_{mean} and EAT (Figure 5A,B). However, a significant quadratic relationship between the relative LI and R_{mean} was observed. The relative LI increased significantly with the R_{mean} from 2015 at Xindu to 2017 at Ezhou, and then decreased slightly from 2017 to 2018 at Ezhou (Figure 5C). In addition, the relative LI and $\triangle$BM showed no significant correlation with the CSR (Figure 5D,H). When referred to the correlations between the relative $\triangle$BM and T_{mean}, EAT and R_{mean}, they displayed the reversed results of the relative LI (Figure 5E–G).

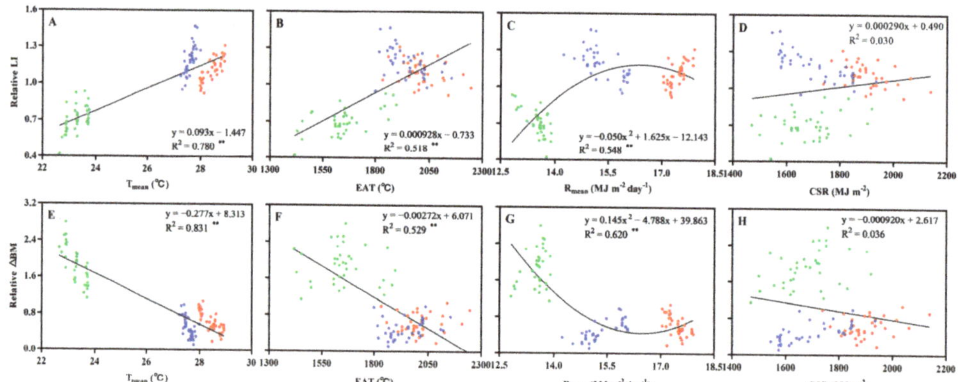

Figure 5. Correlations between the relative LI of the twelve *indica* rice cultivars and T_{mean} (**A**), R_{mean} (**B**), EAT (**C**), and CSR (**D**) on all the sowing dates across two locations and three years. Correlations between the relative $\triangle$BM of the twelve *indica* rice cultivars and T_{mean} (**E**), R_{mean} (**F**), EAT (**G**), and CSR (**H**) on all the sowing dates across two locations and three years. The green, blue, and red points represent the cultivars measured in 2015 at Xindu, and in 2017 and 2018 at Ezhou, respectively. $n = 36$ in each year. **, significant differences at $p < 0.01$.

In order to determine the main temperature and solar radiation parameters that affected the lodging resistance of 12 *indica* rice cultivars, a stepwise regression analysis was conducted (Figure 6). The results of the stepwise regression analysis indicated that only the T_{mean} and R_{mean} had significant effects on the lodging resistance within the temperature and solar radiation parameters. The relative LI increased with the T_{mean} but decreased with the R_{mean} (Figure 6A). The relative $\triangle$BM decreased with the T_{mean} but increased with the R_{mean} (Figure 6B).

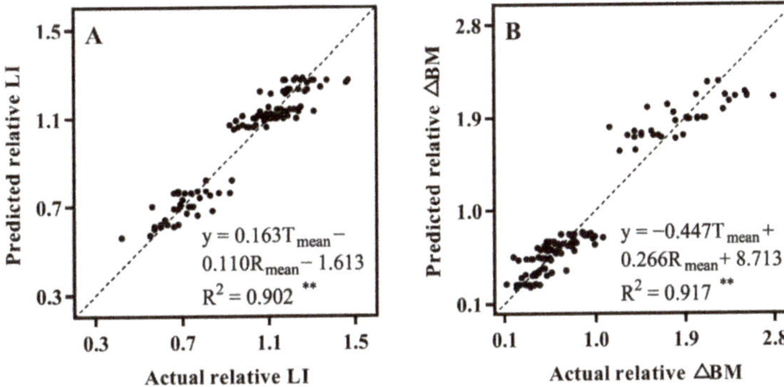

Figure 6. Stepwise regression analysis of the temperature and solar radiation parameters with the relative LI (**A**) and $\triangle$BM (**B**) of twelve *indica* rice cultivars. T_{mean}, average daily temperature. R_{mean}, average daily solar radiation. $n = 108$. **, significant differences at $p < 0.01$.

Following this, the direct and indirect effects of the T_{mean} and R_{mean} on the relative LI and $\triangle$BM were investigated using a path analysis to reveal the relative importance of the T_{mean} and R_{mean} on the lodging resistance of 12 *indica* rice cultivars. The results of the path analysis are shown in Table 4. The direct effect of the T_{mean} on the relative LI was 1.556, and the indirect effect of T_{mean} on the relative LI through the R_{mean} was −0.673. The direct effect of the R_{mean} on the relative LI was −0.759, and the indirect effect of the R_{mean} on the relative LI through the T_{mean} was 1.380. These results indicated that the T_{mean} and R_{mean} independently had a positive and negative effect on the relative LI, respectively. Additionally, the absolute value of the direct effect of the T_{mean} (1.556) was higher than that of the R_{mean} (0.759), and the correlation coefficients of the relative LI with the T_{mean} and R_{mean} were all positive values ($r = 0.883$** and 0.621**, respectively), which implied that the T_{mean} had a greater effect on the lodging resistance than the R_{mean}. The same results could also be obtained from the direct and indirect effects of the T_{mean} and R_{mean} on the relative $\triangle$BM. Thus, the correlations between the R_{mean} and lodging-related traits were not analyzed in the following part.

Table 4. The direct and indirect effects of the T_{mean} and R_{mean} on the relative LI and $\triangle$BM of twelve *indica* rice cultivars.

Dependent Variable	Independent Variable	Correlation Coefficient	Direct Effect	Indirect Effect	
				T_{mean}[a]	R_{mean}[a]
Relative LI	T_{mean}	0.883 ** [b]	1.556	-	−0.673
	R_{mean}	0.621 **	−0.759	1.380	-
Relative $\triangle$BM	T_{mean}	−0.912 **	−1.473	-	0.561
	R_{mean}	−0.674 **	0.633	−1.307	-

[a] T_{mean} and R_{mean} indicate the average daily temperature and average daily solar radiation, respectively.
[b] ** indicate significant differences at $p < 0.01$.

3.4. Correlation analysis of the T_{mean} and Relative Lodging-Related Traits

In order to clarify the lodging-related traits affected by the T_{mean}, the correlations between them were analyzed (Table 5). The relative values of the breaking resistance (BR) and bending moment at breaking (M) were mostly affected by the T_{mean} with the highest coefficient r on the three sowing dates of each year and on all the sowing dates across two locations and three years. Furthermore, the relative value of the stem length (SL) had highly significant positive correlations with the T_{mean} on the three sowing dates of each year and on all the sowing dates across the two locations and three years. However, the rest lodging-related traits showed inconsistent correlations with the T_{mean}. These results implied that the BR, M and SL were the major lodging-related traits affected by the T_{mean}.

Table 5. Analysis of the correlations between the relative value of lodging-related traits of twelve *indica* rice cultivars and the T_{mean} during the determined growth durations on the three sowing dates of different years and on all the sowing dates across two locations and three years.

Year/Location	CL [a]	SL	PL	BR	FW	BM	M
2015/Xindu	0.424 ** [b]	0.708 **	−0.645 **	−0.912 **	−0.365 *	−0.207	−0.912 **
2017/Ezhou	0.024	0.456 **	−0.547 **	−0.603 **	−0.202	−0.148	−0.603 **
2018/Ezhou	−0.303	0.782 **	−0.795 **	−0.882 **	−0.638 **	−0.627 **	−0.882 **
All	0.794 **	0.789 **	−0.014	−0.864 **	−0.273 **	0.195 *	−0.864 **

[a] CL, culm length; SL, stem length; PL, panicle length; BR, breaking resistance; FW, fresh weight; BM, bending moment of the whole plant; M, bending moment at breaking. [b] * and ** indicate significant differences at $p < 0.05$ and $p < 0.01$, respectively.

The M value, which is used to determine the physical strength of the culm, can be further divided into two parts: the section modulus (SM), which is directly influenced by the culm diameter (CD) and culm wall thickness (CT); the bending stress (BS), which is an indicator of culm stiffness. In order to determine the reason behind the significant variation in the M value affected by the T_{mean}, the SM, BS, and other lodging-related traits were measured in 2018 at Ezhou. Additionally, correlation analysis was also conducted between the measured lodging-related traits and the T_{mean} (Table 6). The results of the correlation analysis showed that the T_{mean} had significant negative correlations with the relative value of CD, CT, SM and cellulose content (CC), but displayed significant positive correlations with the relative value of lignin content (LC), total content of lignin and cellulose (TC), and bending stress (BS). In addition, the absolute value of r between the T_{mean} and the relative value of SM was 0.837, which was higher than that between the T_{mean} and the relative value of BS (0.437).

Table 6. Analysis of the correlations between the relative value of lodging-related traits of twelve *indica* rice cultivars and the T_{mean} during the determined growth durations on the three sowing dates in 2018 at Ezhou.

Year/Location	CD [a]	CT	SM	LC	CC	TC	BS
2018/Ezhou	−0.851 ** [b]	−0.754 **	−0.837 **	0.685 **	−0.551 **	0.332 *	0.437 **

[a] CD, culm diameter; CT, culm wall thickness; SM, section modulus; LC, lignin content; CC, cellulose content; TC, total content of lignin and cellulose; BS, bending stress. [b] * and ** indicate significant differences at $p < 0.05$ and $p < 0.01$, respectively.

4. Discussions

As both of the experimental sites have a subtropical monsoon type of climate, the temperatures gradually increase from spring to summer and decrease in autumn (Figure 1). However, Xindu is in the basin area, while Ezhou is in the plain area. Xindu is situated at a higher altitude than Ezhou. The differences in topography and altitude generate different temperature and solar radiation conditions between the two locations (Figure 2). Rice is often sown in the late spring at Xindu and Ezhou, and delayed sowing will make rice suffer from high temperatures sooner. Since high temperatures accelerated the growth of rice and advanced the rice maturity [78,79], the determined growth durations reduced with the delay in the sowing dates (Table 1). Hence, the average daily temperatures (T_{mean}) during the determined growth durations increased with the delayed sowing dates (Figure 2). As a high temperature tended to be accompanied by higher solar radiation [69], the average daily solar radiation (R_{mean}) also increased with the T_{mean} from SD7 to SD9. However, the R_{mean} varied little from SD1 to SD3 and even decreased from SD4 to SD6. This was due to the increase in cloudy and rainy days, which resulted in low sunshine hours and solar radiation. Delayed sowing can enhance the lodging resistance in winter wheat [80]. However, the lodging resistance of *indica* rice cultivars weakened with the delayed sowing dates in our research (Figure 3). Therefore, optimizing sowing dates could be a useful strategy for improving lodging resistance and growing crops to adapt to climate change.

In this study, we found that whether the range of the T_{mean} was less than 1 °C across the three sowing dates in each year or more than 4 °C across the two locations (Figure 2), the increased T_{mean} significantly decreased the lodging resistance of *indica* rice cultivars (Figures 4A,E and 5A,E), which indicated that the lodging resistance was more sensitive to the T_{mean} than the other temperature and solar radiation parameters. This could also be demonstrated by the results of stepwise regression analysis and path analysis. However, the response of the lodging resistance to other crops could be different. Previous research found that high temperatures showed an inconsistent effect on the stem-lodging resistance in canola plants, but reduced the root-lodging resistance significantly [81]. The results of stepwise regression analysis and path analysis also showed that the R_{mean} had a positive effect on the lodging resistance (Figure 5, Table 4). That is to say, an increase in the R_{mean} would improve the lodging resistance. However, this effect was not detectable under the greater negative influence of increased T_{mean} on the lodging resistance across the three sowing dates in 2018 and across the two locations (Figures 4C,G and 5C,G). However, the lodging resistance displayed a slightly increasing trend from 2017 to 2018 at Ezhou (Figure 5C,G). We inferred that the positive effect of the increased R_{mean} on the lodging resistance outweighed the negative effect of the increased T_{mean} because of the larger differences ($\triangle R_{mean}$ = 2.1 MJ m^{-2} d^{-1}) in the R_{mean} than the differences ($\triangle T_{mean}$ = 0.8 °C) in the T_{mean} between 2017 and 2018 at Ezhou (Figure 2). Several researchers found that low solar radiation could promote the vertical elongation of plants, such as an increase in the internode length and plant height, and restrain the lateral growth, such as a decrease in the culm diameter and wall thickness, and reduce cell wall carbohydrates; thus, low solar radiation could cause crop lodging [82–87]. Our results were similar to the previous findings, i.e., the R_{mean} was positively correlated with the lodging resistance. However, under the conditions of our present study, since the T_{mean} had a greater effect than the R_{mean} (Table 4), we could not establish the relationship between solar radiation and lodging-related traits.

According to the correlation analysis between the T_{mean} and lodging-related traits (Table 5), we found that with the increased T_{mean}, it was mainly the reduction in breaking resistance (BR) and the bending moment at breaking (M) of the basal second internode that weakened the lodging resistance of *indica* rice cultivars. The decreased stem-bending strength due to short periods of high temperature stress could also be found in canola [88]. In addition, the stem length (SL) of the basal second internode increased significantly with the increased T_{mean}, which could be a reason for the decline in the BR [89].

The culm stiffness of rice plants, which is expressed by the bending stress (BS), is a product of the structural carbohydrates (primarily lignin and cellulose) and non-structural carbohydrates contents at the lower internode. Carbohydrates in rice culms are accumulated before heading and transported to the ears after heading and it is mainly the non-structural carbohydrates which are transported to grains for grain filling [90,91]. Thus, the culm stiffness increases with more structural carbohydrates in the basal stems [33,41]. Li et al. [92] observed that the cellulose content (CC) of rice *brittle culm1* (*bc1*) was lower than that of the wild type, but the lignin content (LC) was higher. However, the total content of lignin and cellulose (TC) of *bc1* was lower than that of the wild type, which led to the weaker culm strength of *bc1*. In our study, the increase in LC was more than the decrease in CC as the T_{mean} increased. As a result, the TC increased with the T_{mean}, which caused the increased BS (Table 6). The physical strength of the culm, which was represented by the M value, was significantly and positively correlated with the BS and section modulus (SM) [20,32]. Despite the BS increasing with the T_{mean}, the decrease in the SM was even more important with the higher correlation coefficient, which was responsible for the significant reduction in the M value (Table 6). Additionally, the decrease in the SM was attributed to the reduced culm diameter (CD) and culm wall thickness (CT) of the basal second internode with the increased T_{mean}. As a consequence, CD, CT, and SL were the major factors influencing the physical strength of the culm under the effect of temperature.

The soil properties of the experimental fields at Xindu and Ezhou were different, which might affect the lodging resistance of tested rice cultivars. However, rice had been cultivated in the experimental fields for many years, which indicated that the soil properties were suitable for rice growth and development. In addition, the fertilizer applications between the experimental fields at Xindu and Ezhou were identical, resulting in a similar uptake of nitrogen, potassium and silicon from the soil, which caused the soil properties to have little influence on the lodging resistance [93]. In the research of Niu, Feng, Ding and Li [47], strong wind and heavy rain were the two most important causes of lodging. Even though there was no strong wind and heavy rain during the determined growth durations in our study, we selected 12 *indica* rice cultivars which did not lodge in the fields for analysis in order to exclude the influence of wind and rain on the plant lodging. Therefore, ordinary wind and rain had little effect on the lodging resistance of *indica* rice cultivars.

Although the lodging resistance of 12 *indica* rice cultivars significantly weakened with the increased temperature, each of the cultivars responded differently to the temperature (Table 3). The lodging resistance of lodging-moderate cultivar Chuanxiang 29B was most affected by the temperature with the highest coefficient of variation, and that of lodging-resistant cultivar Jiangan responded least to the temperature with the lowest coefficient of variation. This implied that the lodging-resistant *indica* cultivar had the potential to adapt to a higher temperature.

5. Conclusions

The average daily temperature gradually increased by about 1 °C across the three sowing dates in each year and significantly increased by about 5 °C across the two locations. The average solar radiation varied inconsistently across the three years but prominently increased by about 4 MJ m^{-2} day^{-1} across the two locations. Under this condition, the temperature had greater negative effects on the lodging resistance of *indica* rice cultivars than the positive effects of solar radiation. That is to say, temperature was the main factor that affected the lodging resistance of *indica* rice cultivars. With the delay in the sowing dates, the lodging resistance of *indica* rice cultivars showed a significant decrease (LI increased by 22.65%, 19.82% and 17.58% in 2015, 2017 and 2018, respectively; △BM decreased by 35.52%, 57.62% and 47.93% in 2015, 2017 and 2018, respectively) along with an increase in the average daily temperature. Among the morphological traits, the culm diameter and culm wall thickness decreased, and the stem length increased significantly with the increased average daily temperature, which led to the slender basal second internode. As the biochemical trait of the lignin content increased more than the cellulose content decreased, the total content of the lignin and cellulose increased with the increased average daily temperature, which resulted in the increased culm stiffness. In summary, owing to the slender basal second internode, the mechanical trait of the bending moment at breaking decreased significantly with the increased average daily temperature, increasing the potential risk of lodging in *indica* rice cultivars. The slender basal internodes would become a critical reason for the *indica* rice lodging with rising temperatures due to global warming.

Supplementary Materials: The following supporting information can be downloaded at: https://www.mdpi.com/article/10.3390/agronomy12112603/s1, Table S1: The genetic background and the origin of the experimental rice cultivars classified into three different lodging resistance groups; Table S2: Analysis of variance of temperature and solar radiation parameters, LI, and $\triangle$BM.

Author Contributions: Conceptualization, R.Z. and X.W.; Methodology, X.L., Z.W., L.F. and X.W.; Software, X.L.; Formal Analysis, X.L. and Z.W.; Investigation, X.L., Z.W., L.F., Z.Y. and T.L.; Resources, R.Z.; Data Curation, X.L. and Z.W.; Writing—Original Draft Preparation, X.L.; Writing—Review & Editing, Z.D., W.L. and X.W.; Visualization, X.L., Z.D. and X.W.; Supervision, R.Z. and Z.H.; Project Administration, X.W. All authors have read and agreed to the published version of the manuscript.

Funding: This research was funded by the National Key Research and Development Project (2021YFE0101000), the China Agriculture Research System (CARS-01-08), and the Key R&D Program of Hubei Province in China (2021BBA079).

Data Availability Statement: The data presented in this study are available on request from the corresponding author. The data are not publicly available due to their relevance to an ongoing Ph.D. thesis.

Acknowledgments: We thank the staff of the State Key Laboratory of Hybrid Rice, Wuhan University and the Crop Research Institute, Sichuan Academy of Agricultural Science for their experimental support.

Conflicts of Interest: The authors declare no conflict of interest. The funders had no role in the design of the study; in the collection, analyses, or interpretation of data; in the writing of the manuscript; or in the decision to publish the results.

References

1. Lang, Y.; Yang, X.; Wang, M.; Zhu, Q. Effects of lodging at different filling stages on rice yield and grain quality. *Rice Sci.* **2012**, *19*, 315–319. [CrossRef]
2. Hoshikawa, K.; Wang, S. Studies on lodging in rice plants: I. A general observation on lodged rice culms. *Jpn. J. Crop Sci.* **1990**, *59*, 809–814. [CrossRef]
3. Kono, M. Physiological aspects of lodging. *Physiology* **1995**, *2*, 971–982.
4. Nakajima, T.; Yoshida, M.; Tomimura, K. Effect of lodging on the level of mycotoxins in wheat, barley, and rice infected with the Fusarium graminearum species complex. *J. Gen. Plant Pathol.* **2008**, *74*, 289–295. [CrossRef]
5. Khush, G.S. Green revolution: Preparing for the 21st century. *Genome* **1999**, *42*, 646–655. [CrossRef]
6. Khush, G.S. Green revolution: The way forward. *Nat. Rev. Genet.* **2001**, *2*, 815–822. [CrossRef]
7. Hedden, P. The genes of the Green Revolution. *Trends Genet.* **2003**, *19*, 5–9. [CrossRef]
8. Dalrymple, D.G. The Development and Adoption of High-Yielding Varieties of Wheat and Rice in Developing Countries. *Am. J. Agric. Econ.* **1985**, *67*, 1067–1073. [CrossRef]
9. Chen, L.; Yi, Y.; Wang, W.; Zeng, Y.; Tan, X.; Wu, Z.; Chen, X.; Pan, X.; Shi, Q.; Zeng, Y. Innovative furrow ridging fertilization under a mechanical direct seeding system improves the grain yield and lodging resistance of early indica rice in South China. *Field Crops Res.* **2021**, *270*, 108184. [CrossRef]
10. Wang, W.; Du, J.; Zhou, Y.; Zeng, Y.; Tan, X.; Pan, X.; Shi, Q.; Wu, Z.; Zeng, Y. Effects of different mechanical direct seeding methods on grain yield and lodging resistance of early indica rice in South China. *J. Integr. Agric.* **2021**, *20*, 1204–1215. [CrossRef]
11. Khush, G.S. Breaking the yield frontier of rice. *GeoJournal* **1995**, *35*, 329–332. [CrossRef]
12. Berry, P.M.; Sterling, M.; Spink, J.H.; Baker, C.J.; Sylvester-Bradley, R.; Mooney, S.J.; Tams, A.R.; Ennos, A.R. Understanding and reducing lodging in cereals. *Adv. Agron.* **2004**, *84*, 215–269. [CrossRef]
13. Berry, P.M.; Sterling, M.; Mooney, S.J. Development of a model of lodging for barley. *J. Agron. Crop Sci.* **2006**, *192*, 151–158. [CrossRef]
14. Berry, P.M.; Sylvester-Bradley, R.; Berry, S. Ideotype design for lodging-resistant wheat. *Euphytica* **2007**, *154*, 165–179. [CrossRef]
15. Mulder, E.G. Effect of mineral nutrition on lodging of cereals. *Plant Soil* **1954**, *5*, 246–306. [CrossRef]
16. Hirano, K.; Ordonio, R.L.; Matsuoka, M. Engineering the lodging resistance mechanism of post-Green Revolution rice to meet future demands. *Proc. Jpn. Acad. B-Phys.* **2017**, *93*, 220–233. [CrossRef]
17. Islam, M.S.; Peng, S.B.; Visperas, R.M.; Ereful, N.; Bhuiya, M.S.U.; Julfiquar, A.W. Lodging-related morphological traits of hybrid rice in a tropical irrigated ecosystem. *Field Crops Res.* **2007**, *101*, 240–248. [CrossRef]
18. Zhu, C.; Ziska, L.H.; Sakai, H.; Zhu, J.; Hasegawa, T. Vulnerability of lodging risk to elevated CO2 and increased soil temperature differs between rice cultivars. *Eur. J. Agron.* **2013**, *46*, 20–24. [CrossRef]
19. Zhang, W.; Li, G.; Yang, Y.; Li, Q.; Zhang, J.; Liu, J.; Wang, S.; Tang, S.; Ding, Y. Effects of nitrogen application rate and ratio on lodging resistance of super rice with different genotypes. *J. Integr. Agric.* **2014**, *13*, 63–72. [CrossRef]
20. Zhang, W.; Wu, L.; Ding, Y.; Weng, F.; Wu, X.; Li, G.; Liu, Z.; Tang, S.; Ding, C.; Wang, S. Top-dressing nitrogen fertilizer rate contributes to decrease culm physical strength by reducing structural carbohydrate content in japonica rice. *J. Integr. Agric.* **2016**, *15*, 992–1004. [CrossRef]

21. Guo, Z.; Liu, X.; Zhang, B.; Yuan, X.; Xing, Y.; Liu, H.; Luo, L.; Chen, G.; Xiong, L. Genetic analyses of lodging resistance and yield provide insights into post-Green-Revolution breeding in rice. *Plant Biotechnol. J.* **2021**, *19*, 814–829. [CrossRef]

22. Hossain, K.A.; Horiuchi, T.; Miyagawa, S. Effects of powdered rice chaff application on Si and N absorption, lodging resistance and yield in rice plants (*Oryza sativa* L.). *Plant Prod. Sci.* **1999**, *2*, 159–164. [CrossRef]

23. Duy, P.Q.; Abe, A.; Hirano, M.; Sagawa, S.; Kuroda, E. Analysis of lodging-resistant characteristics of different rice genotypes grown under the standard and nitrogen-free basal dressing accompanied with sparse planting density practices. *Plant Prod. Sci.* **2004**, *7*, 243–251. [CrossRef]

24. Mobasser, H.R.; Yadi, R.; Azizi, M.; Ghanbari, A.M.; Samdaliri, M. Effect of density on morphological characteristics related-lodging on yield and yield components in varieties rice (*Oryza sativa* L.) In Iran. *Am.-Eurasian J. Agric. Environ. Sci.* **2009**, *5*, 745–754.

25. Luo, X.Y.; Wu, Z.F.; Fu, L.; Dan, Z.W.; Yuan, Z.Q.; Liang, T.; Zhu, R.S.; Hu, Z.L.; Wu, X.T. Evaluation of lodging resistance in rice based on an optimized parameter from lodging index. *Crop Sci.* **2022**, *62*, 1318–1332. [CrossRef]

26. Matsuda, T.; Kawahara, H.; Chonan, N. Histological studies on breaking resistance of lower internodes in rice culm: IV. The roles of each tissue of internode and leaf sheath in breaking resistance. *Jpn. J. Crop Sci.* **1983**, *52*, 355–361. [CrossRef]

27. Yagi, T. Studies on breeding for culm stifness in rice: 1. Varietal differences in culm stiffness and its related traits. *Jpn. J. Breed.* **1983**, *33*, 411–422. [CrossRef]

28. Ichii, M.; Hada, K. Application of ratoon to a test of agronomic characters in rice breeding. II. The relationship between ratoon ability and lodging resistance. *Jpn. J. Breed.* **1983**, *33*, 251–258. [CrossRef]

29. Ookawa, T.; Ishihara, K. Varietal difference of physical characteristics of the culm related to lodging resistance in paddy rice. *Jpn. J. Crop Sci.* **1992**, *61*, 419–425. [CrossRef]

30. Kashiwagi, T.; Togawa, E.; Hirotsu, N.; Ishimaru, K. Improvement of lodging resistance with QTLs for stem diameter in rice (*Oryza sativa* L.). *Theor. Appl. Genet.* **2008**, *117*, 749–757. [CrossRef]

31. Li, H.; Zhang, X.; Li, W.; Xu, Z.; Xu, H. Lodging resistance in japonica rice varieties with different panicle types. *Chin. J. Rice Sci.* **2009**, *23*, 191–196. [CrossRef]

32. Ookawa, T.; Yasuda, K.; Kato, H.; Sakai, M.; Seto, M.; Sunaga, K.; Motobayashi, T.; Tojo, S.; Hirasawa, T. Biomass production and lodging resistance in 'Leaf Star', a new long-culm rice forage cultivar. *Plant Prod. Sci.* **2010**, *13*, 58–66. [CrossRef]

33. Zhang, J.; Li, G.; Song, Y.; Liu, Z.; Yang, C.; Tang, S.; Zheng, C.; Wang, S.; Ding, Y. Lodging resistance characteristics of high-yielding rice populations. *Field Crops Res.* **2014**, *161*, 64–74. [CrossRef]

34. Fan, C.; Li, Y.; Hu, Z.; Hu, H.; Wang, G.; Li, A.; Wang, Y.; Tu, Y.; Xia, T.; Peng, L.; et al. Ectopic expression of a novel OsExtensin-like gene consistently enhances plant lodging resistance by regulating cell elongation and cell wall thickening in rice. *Plant Biotechnol. J.* **2018**, *16*, 254–263. [CrossRef] [PubMed]

35. Hirano, K.; Okuno, A.; Hobo, T.; Ordonio, R.; Shinozaki, Y.; Asano, K.; Kitano, H.; Matsuoka, M. Utilization of stiff culm trait of rice smos1 mutant for increased lodging resistance. *PLoS ONE* **2014**, *9*, e96009. [CrossRef]

36. Tanaka, K.; Murata, K.; Yamazaki, M.; Onosato, K.; Miyao, A.; Hirochika, H. Three distinct rice cellulose synthase catalytic subunit genes required for cellulose synthesis in the secondary wall. *Plant Physiol.* **2003**, *133*, 73–83. [CrossRef]

37. Li, Q.; Fu, C.; Liang, C.; Ni, X.; Zhao, X.; Chen, M.; Ou, L. Crop Lodging and The Roles of Lignin, Cellulose, and Hemicellulose in Lodging Resistance. *Agronomy* **2022**, *12*, 1795. [CrossRef]

38. Ishimaru, K.; Togawa, E.; Ookawa, T.; Kashiwagi, T.; Madoka, Y.; Hirotsu, N. New target for rice lodging resistance and its effect in a typhoon. *Planta* **2008**, *227*, 601–609. [CrossRef] [PubMed]

39. Li, X.J.; Yang, Y.; Yao, J.L.; Chen, G.X.; Li, X.H.; Zhang, Q.F.; Wu, C.Y. FLEXIBLE CULM 1 encoding a cinnamyl-alcohol dehydrogenase controls culm mechanical strength in rice. *Plant Mol. Biol.* **2009**, *69*, 685–697. [CrossRef]

40. Zhang, F.; Jin, Z.; Ma, G.; Shang, W.; Liu, H.; Xu, M.; Liu, Y. Relationship between lodging resistance and chemical contents in culms and sheaths of japonica rice during grain filling. *Rice Sci.* **2010**, *17*, 311–318. [CrossRef]

41. Zhang, J.; Li, G.; Huang, Q.; Liu, Z.; Ding, C.; Tang, S.; Chen, L.; Wang, S.; Ding, Y.; Zhang, W. Effects of culm carbohydrate partitioning on basal stem strength in a high-yielding rice population. *Crop J.* **2017**, *5*, 478–487. [CrossRef]

42. Yang, S.; Xie, L.; Zheng, S.; Li, J.; Yuan, J. Effects of nitrogen rate and transplanting density on physical and chemical characteristics and lodging resistance of culms in hybrid rice. *Acta Agron. Sin.* **2009**, *35*, 93–103. [CrossRef]

43. Wu, W.; Huang, J.; Cui, K.; Nie, L.; Wang, Q.; Yang, F.; Shah, F.; Yao, F.; Peng, S. Sheath blight reduces stem breaking resistance and increases lodging susceptibility of rice plants. *Field Crops Res.* **2012**, *128*, 101–108. [CrossRef]

44. Gong, D.; Zhang, X.; Yao, J.; Dai, G.; Yu, G.; Zhu, Q.; Gao, Q.; Zheng, W. Synergistic effects of bast fiber seedling film and nano-silicon fertilizer to increase the lodging resistance and yield of rice. *Sci. Rep.-UK* **2021**, *11*, 12788. [CrossRef]

45. Hong, W.; Chen, Y.; Huang, S.; Li, Y.; Wang, Z.; Tang, X.; Pan, S.; Tian, H.; Mo, Z. Optimization of nitrogen–silicon (N-Si) fertilization for grain yield and lodging resistance of early-season indica fragrant rice under different planting methods. *Eur. J. Agron.* **2022**, *136*, 126508. [CrossRef]

46. Ohe, M.; Tamura, A.; Mimoto, H. Effects of deep water treatment on the growth of culms and the lodging resistance in Japonica type paddy rice (*Oryza sativa* L.) cultivars. *Jpn. J. Crop Sci.* **1996**, *65*, 238–244. [CrossRef]

47. Niu, L.; Feng, S.; Ding, W.; Li, G. Influence of speed and rainfall on large-scale wheat lodging from 2007 to 2014 in China. *PLoS ONE* **2016**, *11*, e0157677. [CrossRef]

48. Shimono, H.; Okada, M.; Yamakawa, Y.; Nakamura, H.; Kobayashi, K.; Hasegawa, T. Lodging in rice can be alleviated by atmospheric CO_2 enrichment. *Agric. Ecosyst. Environ.* **2007**, *118*, 223–230. [CrossRef]

49. Revilla-Molina, I.M.; Bastiaans, L.; Van Keulen, H.; Kropff, M.J.; Hui, F.; Castilla, N.P.; Mew, T.W.; Zhu, Y.Y.; Leung, H. Does resource complementarity or prevention of lodging contribute to the increased productivity of rice varietal mixtures in Yunnan, China? *Field Crops Res.* **2009**, *111*, 303–307. [CrossRef]
50. Olagunju, S.O.; Atayese, M.O.; Sakariyawo, O.S.; Dare, E.O.; Tang, C. Effects of multi-growth stage water deficit and orthosilicic acid fertiliser on lodging resistance of rice cultivars. *Crop Pasture Sci.* **2022**, *73*, 370–389. [CrossRef]
51. Schneider, S.H. What is 'dangerous' climate change? *Nature* **2001**, *411*, 17–19. [CrossRef] [PubMed]
52. Yuan, M.; Leirvik, T.; Wild, M. Global trends in downward surface solar radiation from spatial interpolated ground observations during 1961–2019. *J. Clim.* **2021**, *34*, 9501–9521. [CrossRef]
53. Ramanathan, V.; Feng, Y. Air pollution, greenhouse gases and climate change: Global and regional perspectives. *Atmos. Environ.* **2009**, *43*, 37–50. [CrossRef]
54. Shakun, J.D.; Clark, P.U.; He, F.; Marcott, S.A.; Mix, A.C.; Liu, Z.Y.; Otto-Bliesner, B.; Schmittner, A.; Bard, E. Global warming preceded by increasing carbon dioxide concentrations during the last deglaciation. *Nature* **2012**, *484*, 49–54. [CrossRef] [PubMed]
55. IPCC (Intergovernmental Panel on Climate Change). *Climate Change 2014: Synthesis Report. Contribution of Working Groups I, II and III to the Fourth Assessment Report of the Intergovernmental Panel on Climate Change*; IPCC: Geneva, Switzerland, 2014.
56. Stanhill, G.; Cohen, S. Global dimming: A review of the evidence for a widespread and significant reduction in global radiation with discussion of its probable causes and possible agricultural consequences. *Agric. For. Meteorol.* **2001**, *107*, 255–278. [CrossRef]
57. Kambezidis, H.D.; Kaskaoutis, D.G.; Kharol, S.K.; Moorthy, K.K.; Satheesh, S.K.; Kalapureddy, M.C.R.; Badarinath, K.V.S.; Sharma, A.R.; Wild, M. Multi-decadal variation of the net downward shortwave radiation over south Asia: The solar dimming effect. *Atmos. Environ.* **2012**, *50*, 360–372. [CrossRef]
58. Zhou, N.-B.; Zhang, J.; Fang, S.-L.; Wei, H.-Y.; Zhang, H.-C. Effects of temperature and solar radiation on yield of good eating-quality rice in the lower reaches of the Huai River Basin, China. *J. Integr. Agric.* **2021**, *20*, 1762–1774. [CrossRef]
59. Deng, F.; Zhang, C.; He, L.; Liao, S.; Li, Q.; Li, B.; Zhu, S.; Gao, Y.; Tao, Y.; Zhou, W.; et al. Delayed sowing date improves the quality of mechanically transplanted rice by optimizing temperature conditions during growth season. *Field Crops Res.* **2022**, *281*, 108493. [CrossRef]
60. Katsura, K.; Maeda, S.; Lubis, I.; Horie, T.; Cao, W.; Shiraiwa, T. The high yield of irrigated rice in Yunnan, China. *Field Crops Res.* **2008**, *107*, 1–11. [CrossRef]
61. Tao, F.; Zhang, Z.; Shi, W.; Liu, Y.; Xiao, D.; Zhang, S.; Zhu, Z.; Wang, M.; Liu, F. Single rice growth period was prolonged by cultivars shifts, but yield was damaged by climate change during 1981-2009 in China, and late rice was just opposite. *Glob. Chang. Biol.* **2013**, *19*, 3200–3209. [CrossRef]
62. Weng, F.; Zhang, W.; Wu, X.; Xu, X.; Ding, Y.; Li, G.; Liu, Z.; Wang, S. Impact of low-temperature, overcast and rainy weather during the reproductive growth stage on lodging resistance of rice. *Sci. Rep.-UK* **2017**, *7*, 46596. [CrossRef] [PubMed]
63. Agrama, H.A.; Yan, W.G.; Lee, F.; Fjellstrom, R.; Chen, M.H.; Jia, M.; McClung, A. Genetic Assessment of a Mini-Core Subset Developed from the USDA Rice Genebank. *Crop Sci.* **2009**, *49*, 1336–1346. [CrossRef]
64. China Meteorological Data Service Centre. Available online: http://data.cma.cn/en (accessed on 18 December 2019).
65. Angstrom, A. Solar and terrestrial radiation. *Quart. J. Roy. Met. Soc.* **1924**, *50*, 121–126. [CrossRef]
66. Hussain, M.; Rahman, L.; Rahman, M.M. Techniques to obtain improved predictions of global radiation from sunshine duration. *Renew. Energy* **1999**, *18*, 263–275. [CrossRef]
67. Chen, R.; Ersi, K.; Yang, J.; Lu, S.; Zhao, W. Validation of five global radiation models with measured daily data in China. *Energy Convers. Manag.* **2004**, *45*, 1759–1769. [CrossRef]
68. Nash, J.E.; Sutcliffe, J.V. River flow forecasting through conceptual models part I—A discussion of principles. *J. Contam. Hydrol.* **1970**, *10*, 282–290. [CrossRef]
69. Deng, N.; Ling, X.; Sun, Y.; Zhang, C.; Fahad, S.; Peng, S.; Cui, K.; Nie, L.; Huang, J. Influence of temperature and solar radiation on grain yield and quality in irrigated rice system. *Eur. J. Agron.* **2015**, *64*, 37–46. [CrossRef]
70. Seko, H. Studies on lodging in rice plants. *Bull. Kyusyu Agric. Exp. Sta.* **1962**, *7*, 419–499.
71. Ookawa, T.; Ishihara, K. Varietal difference of the cell wall components affecting the bending stress of the culm in relation to the lodging resistance in paddy rice. *Jpn. J. Crop Sci.* **1993**, *62*, 378–384. [CrossRef]
72. Ookawa, T.; Hobo, T.; Yano, M.; Murata, K.; Ando, T.; Miura, H.; Asano, K.; Ochiai, Y.; Ikeda, M.; Nishitani, R.; et al. New approach for rice improvement using a pleiotropic QTL gene for lodging resistance and yield. *Nat. Commun.* **2010**, *1*, 132. [CrossRef]
73. Updegraff, D.M. Semimicro determination of cellulose inbiological materials. *Anal. Biochem.* **1969**, *32*, 420–424. [CrossRef]
74. Fukushima, R.S.; Hatfield, R.D. Comparison of the acetyl bromide spectrophotometric method with other analytical lignin methods for determining lignin concentration in forage samples. *J. Agric. Food Chem.* **2004**, *52*, 3713–3720. [CrossRef] [PubMed]
75. Johnson, D.B.; Moore, W.E.; Zank, L.C. The spectrophotometric determination of lignin in small wood samples. *Tappi* **1961**, *44*, 793–798.
76. Fan, H.; Liu, X. Comparison and optimization of various non-dimensionalized methods based on comprehensive evaluation method–a case study of land development in Yongdeng county of Lanzhou city. *Hunan Agric. Sci.* **2010**, *17*, 163–166. [CrossRef]
77. Luo, H.F.; Zhang, J.Y.; Jia, W.J.; Ji, F.M.; Yan, Q.; Xu, Q.; Ke, S.; Ke, J.S. Analyzing the role of soil and rice cadmium pollution on human renal dysfunction by correlation and path analysis. *Environ. Sci. Pollut. Res. Int.* **2017**, *24*, 2047–2054. [CrossRef]
78. Zhang, T.; Huang, Y.; Yang, X. Climate warming over the past three decades has shortened rice growth duration in China and cultivar shifts have further accelerated the process for late rice. *Global Change Biol.* **2013**, *19*, 563–570. [CrossRef]

79. Guo, Y.; Wu, W.; Du, M.; Liu, X.; Wang, J.; Bryant, C.R. Modeling Climate Change Impacts on Rice Growth and Yield under Global Warming of 1.5 and 2.0 °C in the Pearl River Delta, China. *Atmosphere* **2019**, *10*, 567. [CrossRef]
80. Dai, X.; Wang, Y.; Dong, X.; Qian, T.; Yin, L.; Dong, S.; Chu, J.; He, M. Delayed sowing can increase lodging resistance while maintaining grain yield and nitrogen use efficiency in winter wheat. *Crop J.* **2017**, *5*, 541–552. [CrossRef]
81. Wu, W.; Ma, B.L. Assessment of canola crop lodging under elevated temperatures for adaptation to climate change. *Agric. For. Meteorol.* **2018**, *248*, 329–338. [CrossRef]
82. Kamiji, Y.; Hayashi, S.; Horie, T. Influences of nitrogen nutrient and solar radiation in the canopy on length of lower internodes of rice. *Jpn. J. Crop Sci.* **1993**, *62*, 164–171. [CrossRef]
83. Ookawa, T.; Todokoro, Y.; Ishihara, K. Changes in physical and chemical characteristics of culm associated with lodging resistance in paddy rice under different growth conditions and varietal difference of their changes. *Jpn. J. Crop Sci.* **1993**, *62*, 525–533. [CrossRef]
84. Liu, H.; Yang, C.; Li, L. Shade-induced stem elongation in rice seedlings: Implication of tissue-specific phytohormone regulation. *J. Integr. Plant Biol.* **2016**, *58*, 614–617. [CrossRef] [PubMed]
85. Wu, L.; Zhang, W.; Ding, Y.; Zhang, J.; Cambula, E.D.; Weng, F.; Liu, Z.; Ding, C.; Tang, S.; Chen, L.; et al. Shading contributes to the reduction of stem mechanical strength by decreasing cell wall synthesis in japonica rice (*Oryza sativa* L.). *Front. Plant Sci.* **2017**, *8*, 881. [CrossRef] [PubMed]
86. Hussain, S.; Iqbal, N.; Pang, T.; Khan, M.N.; Liu, W.G.; Yang, W.Y. Weak stem under shade reveals the lignin reduction behavior. *J. Integr. Agric.* **2019**, *18*, 496–505. [CrossRef]
87. Zhang, W.; Duan, X.; Yao, X.; Liu, Q.; Xiao, R.; Zhang, X.; Tang, Y.; Wen, M.; Li, J. Effects of shading on stem morphological traits and lodging resistance in heavy type panicle of indica rice. *Chin. Rice* **2020**, *26*, 9–13. [CrossRef]
88. Wu, W.; Shah, F.; Duncan, R.W.; Ma, B.L. Grain yield, root growth habit and lodging of eight oilseed rape genotypes in response to a short period of heat stress during flowering. *Agric. For. Meteorol.* **2020**, *287*, 107954. [CrossRef]
89. Liu, C.; Zheng, S.; Gui, J.; Fu, C.; Yu, H.; Song, D.; Shen, J.; Qin, P.; Liu, X.; Han, B.; et al. Shortened basal internodes encodes a gibberellin 2-oxidase and contributes to lodging resistance in rice. *Mol. Plant* **2018**, *11*, 288–299. [CrossRef]
90. Okawa, S.; Makino, A.; Mae, T. Effect of irradiance on the partitioning of assimilated carbon during the early phase of grain filling in rice. *Ann. Bot.* **2003**, *92*, 357–364. [CrossRef]
91. Slewinski, T.L. Non-structural carbohydrate partitioning in grass stems: A target to increase yield stability, stress tolerance, and biofuel production. *J. Exp. Bot.* **2012**, *63*, 4647–4670. [CrossRef]
92. Li, Y.; Qian, Q.; Zhou, Y.; Yan, M.; Sun, L.; Zhang, M.; Fu, Z.; Wang, Y.; Han, B.; Pang, X.; et al. BRITTLE CULM1, which encodes a COBRA-like protein, affects the mechanical properties of rice plants. *Plant Cell* **2003**, *15*, 2020–2031. [CrossRef]
93. Zhang, S.; Yang, Y.; Zhai, W.; Tong, Z.; Shen, T.; Li, Y.C.; Zhang, M.; Sigua, G.C.; Chen, J.; Ding, F. Controlled-release nitrogen fertilizer improved lodging resistance and potassium and silicon uptake of direct-seeded rice. *Crop Sci.* **2019**, *59*, 2733–2740. [CrossRef]

Article

Orderly Mechanical Seedling-Throwing: An Efficient and High Yielding Establishment Method for Rice Production

Weiqin Wang [†]**, Li Xiang** [†]**, Huabin Zheng and Qiyuan Tang** *****

College of Agronomy, Hunan Agricultural University, Changsha 410128, China
***** Correspondence: qytang@hunau.edu.cn
[†] These authors contributed equally to this work.

Abstract: Developing an efficient and high-yielding mechanical rice establishment system is one of the most important approaches for intensive and large-scale rice production. Recently, an orderly mechanical rice seedling throwing system (OMST) was successfully developed; however, the performance of this system is unknown. In the present study, a two-year field experiment was carried out in a split-plot design with three establishment methods arranged in the main plots, and two elite rice cultivars arranged in the sub-plots. The grain yield and growth-related traits were then determined. The results showed that the grain yield of OMST was significantly higher than manual seedling throwing, and was equivalent to that of manual transplanting, which was mainly due to the variances in panicle number and total spikelet number. Further analysis suggested that the orderly mechanical seedling throwing takes advantage of higher biomass accumulation after heading, increased leaf area index and decreased leaf senescence rate against manual seedling throwing, and more tillers and biomass accumulation at vegetative growth stage as compared to manual transplanting. The present study showed that the OMST is an efficient and high-yielding rice establishment method that may be a promising option to replace traditional manual seedling throwing in rice production.

Keywords: rice establishment method; orderly mechanical seedling throwing; manual seedling throwing; manual transplanting

Citation: Wang, W.; Xiang, L.; Zheng, H.; Tang, Q. Orderly Mechanical Seedling-Throwing: An Efficient and High Yielding Establishment Method for Rice Production. *Agronomy* **2022**, *12*, 2837. https://doi.org/10.3390/agronomy12112837

Academic Editors: Yuxian Zhu, Shaoqing Li and Guangcun He

Received: 30 September 2022
Accepted: 10 November 2022
Published: 13 November 2022

Publisher's Note: MDPI stays neutral with regard to jurisdictional claims in published maps and institutional affiliations.

1. Introduction

Rice is the dominant staple food in China. In recent years, the labor scarcity in rural areas and the increase in rice production costs resulting from rapid urbanization and industrialization has greatly challenged rice production in China [1]. Meanwhile, continuous population growth requires a further improvement in both rice yield and total rice production. It has been estimated that China will need to produce approximately 20% more rice by 2030 to meet its domestic needs [2]. Therefore, the systematic upgrading of rice production from the manual, high input and low-profit method to a simplified, mechanized and efficient system is desperately needed. Over the last decades, a variety of approaches have been adopted in Chinese rice production system, including the development of resource-efficient rice cropping systems [3,4], the promotion of tillage and harvest machines [5], and innovation in precise fertilizer management [6], which have greatly mitigated the disparity between rice food demand and the scarcity in labor and resources. However, the labor demand and the production costs during the rice establishment process are still relatively high and have become a major constraint for rice production in China [7,8]. To solve this problem, simplified and mechanized rice establishment methods need to be addressed.

In traditional rice production, the rice seeds are sown on a nursery to raise rice seedlings, and then the rice seedlings are manually transplanted to the puddled field, which requires high labor input and is no longer suitable for large-scale rice production in China [9]. In recent years, several simplified or mechanized rice establishment methods, such as rice direct-seeding (DSR) and machine transplanting, have been extended rapidly to replace manual transplanting (MT), and the characteristics of these establishment methods have also been intensively studied [9–12]. Meanwhile, seedling throwing (MST) is a simplified rice establishment method in which rice seedlings with soil on their roots are thrown by hand into fields [13]. In comparison with MT, MST greatly reduces the labor input during transplanting, which increases the economic benefits. Significant variances in growth performance between MT and MST were observed. The setback caused by uprooting and transplanting was either smaller or non-existent in MST than that in MT, which may shorten the growth duration and is regarded as an ideal trait for multiple rice cropping where the rice growing season is strictly limited. Moreover, higher tiller generation, vigorous root growth and more panicle numbers were observed in MST system [14,15]. However, unlike the MT system where rice seedlings are transplanted in lines and rows, the rice seedlings in the MST system are randomly and unevenly distributed in the field. Such traits may reduce the ventilation ability and increase the relative air humidity of the MST population especially under high nitrogen input and high planting density condition, which increases the risk of pest/disease infection and lodging and thus lead to a decrease in grain yield. In addition, although the rice transplanting machine has been developed and adopted widely in China, little attention has been paid to the mechanical seedling throwing system.

Recently, an orderly mechanical rice seedling throwing system (OMST) was developed using a seedling-throwing machine (manufactured by Zoomlion Co., Ltd., Changsha, China). The OMST throws the rice seedlings to the field by machine but the seedlings distribution is similar with MT arranged in lines and rows. The OMST has been rapidly promoted in several provinces of China including Hunan, Anhui, Jiangxi, and its advantages in economic profits have been proved in farmers' practices. However, the grain yield performance and the growth traits of the OMST are unknown. It can be speculated that OMST might combine the advantages of MST regarding seedling vigor and fast tiller generation, as well as the advantages of MT regarding orderly population distribution. To test this hypothesis, a two-year field experiment was carried out with three rice establishment systems incorporated, namely orderly mechanical seedling throwing (OMST), manual transplanting (MT) and manual seedling throwing (MST) (Figure 1). The objectives of the present study were to compare the variances between OMST, MT and MST in grain yield, yield components and growth characters, and to determine the extension potential of OMST system in China.

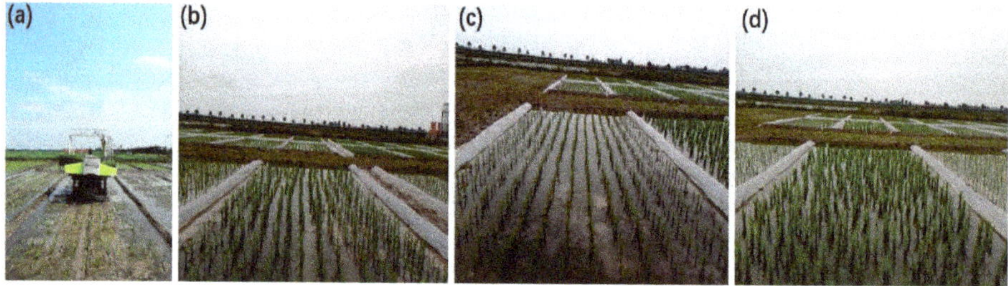

Figure 1. Pictorial illustration of the field performance of orderly mechanical seedling transplanting (**a**,**b**); manual transplanting (**c**); and manual seedling throwing (**d**).

2. Materials and Methods

2.1. Site Description

The present study was conducted at Qianshanhong Town (29°19′ N, 112°15′ E), Yiyang county, Hunan Province, China during 2021 and 2022 growing season (Figure 2). Before the experiment, the soil from the 0–20 cm layer was randomly collected and the soil nutrients content were then determined by the automatic Kjeldahl apparatus method (nitrogen), microwave digestion-flame photometry method (phosphorus and potassium), potassium dichromate volumetric method (organic matter). The soil pH was determined by an automatic soil pH meter (FieldScout PH400, manufactured by SPECTRUM Techno. Inc., Haltom City, TX, USA) The soil was clay loam with soil pH = 6.77. The available nitrogen, phosphorus, potassium, and organic matter were 117.37 mg kg^{-1}, 15.57 mg kg^{-1}, 111.07 mg kg^{-1} and 27.9 g kg^{-1}, respectively. The temperature (daily maximum, daily average, and daily minimum) and rainfall during the rice-growing season at the experimental sites in 2021 and 2022 are shown in Figure 3.

Figure 2. The location of the experimental site.

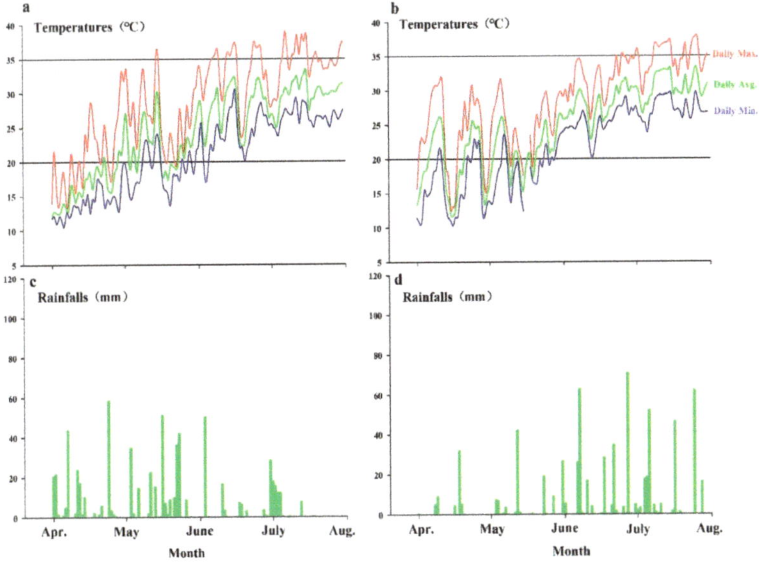

Figure 3. Temperature (daily maximum, daily average and daily minimum) and rainfall during the rice-growing season at the experimental sites in 2021 (**a,c**); and 2022 (**b,d**).

2.2. Experimental Design

The experiment was arranged in randomized complete block design under split plot arrangement with three replications. Three different rice establishment methods, viz., orderly mechanical seedling throwing (OMST), manual transplanting (MT), and manual seed-ling throwing (MST) were arranged in the main plot, and two elite inbred varieties that are widely adopted by local farmers, Xiang-zaoxian24 (XZX24) and Zhongzao39 (ZZ39) were arranged in the sub-plot.

In MT plots, the rice seeds were sown in nursery bed and 25-day-old seedlings were transplanted at a hill spacing of 25 cm × 12 cm with 4 seedlings per hill. In MST and OMST plots, the rice seeds were sown into the holes of the specialized plastic seedling trays with 4 seeds per hole. After covering mud on the surface, the plastic trays were transferred to nursery bed. At 25 days after sowing, the seedlings for MST were manually thrown into the plot at the planting density of 33 hills per m^2, while the seedlings for OMST were mechanically thrown to the plots at hill spacing of 25 cm × 12 cm. Before transplanting, the seedling numbers per hole in the plastic trays were carefully checked and adjusted to ensure the same seedling numbers in each plot. The rice seeds for all the plots were sown on 24 March and transplanted on 18 April in both 2021 and 2022. A fertilizer dose of 150:32:124 of N:P:K kg ha^{-1} was equally applied to each plot. All the P, one third of the N, and half of the K were applied as a basal starter dose, while the residual N was equally split at the middle tillering stage and the panicle initiation stage, and the other 50% of the potassium was top-dressed during panicle initiation. Both basal and top-dressing fertilizers were manually broadcasted to each plot. All the plots were flood irrigated immediately after transplanting at the water depth of 1–3 cm, and at 7 days before harvesting, the field was drained. At early tillering, 25% cyhalofop-butyl was spray-applied to control weed infection. 20% abamectin, 10% imidacloprid, 20% tricyclazole and 5% validamycin were sprayed applied at middle tillering and heading to prevent pest and disease infection.

2.3. Observations

2.3.1. Grain Yield and Yield Components

At maturity, 15 plants near the center of each plot were sampled to determine yield components, aboveground total biomass. Panicle number was counted in each sample to determine the panicle number per m^2, and then plants were separated into straw and panicles. Through hand-threshing, all spikelets parted from the rachis were submerged into the tap water to separate the filled grains from the others. Further screening was completed by a winnowing cleanliness instrument (FJ-1; China Rice Research Institute, Hangzhou, China) to separate the half-filled spikelets from the unfilled spikelets. Three sub-samples of 30.0 g of filled spikelets, 2.0 g of unfilled spikelets, and all of the half-filled spikelets were taken to count the number of spikelets. After oven-drying at 70 °C to constant weight, dry weight of straw, rachis and filled, half-filled and unfilled spikelets were determined. Grain yield was determined from a 5 m^2 area in each plot and adjusted to the standard moisture content of 0.14 g H_2O g^{-1} fresh weight.

2.3.2. Growth Analysis

Aboveground biomass was determined at middle tillering, panicle initiation, heading and maturity stage by collecting samples of 12 plants. Samples oven-drying at 70 °C to constant weight. Leaf area index (LAI, which was calculated as the total leaf area of the plants divide the land area that growth the plants) was assessed manually in 2022 at middle tillering, panicle initiation and heading stage. The leaf area was measured as the length of green leaf (from leaf base to leaf tip) multiplied by the maximum width of the blade and an empirical shape factor (0.75). A chlorophyll meter (SPAD-502, Manufactured by Konica Minolta Ltd. Tokyo, Japan) was used to obtain SPAD values of intact flag leaf at heading and maturity. Three SPAD readings were taken around the midpoint of each leaf blade. 15 SPAD readings from the main stem of the five plants that near the plot center were averaged to represent the mean SPAD value of each plot.

2.3.3. Tiller Numbers

Excluding the three border plants, 10 hills were labeled in each plot to count tillers at fixed intervals from 12 days after transplanting to 54 days after transplanting.

2.4. Statistical Analysis

Data were analyzed by analysis of variance using Statistix 9.0 (manufactured by Analytical Software, Tallahassee, FL, USA). The differences between treatments were separated using the least significance difference (LSD) test at the 0.05 probability level. Graphical representation of the data was performed using Sigmaplot 12.5 (manufactured by Systat Software, Inc., Palo Alto, CA, USA).

3. Results

3.1. Grain Yield and Yield Components

The grain yield of OMST were similar or higher than MT and was significantly increased compared with MST (except for XZX24 in 2022 where the yield difference did not reach significant level) (Figure 4). Averaging across years, the grain yield observed in OMST system was 7.17 and 7.18 t ha^{-1}, which was increased by 16.1% and 15.3% for XZX24 and ZZ39, respectively, as compared with MST. When compared with MT, the grain yield of OMST showed no significant difference for XZX24 in 2021, and for both varieties in 2022. However, for XZX24 in 2021, OMST significantly increased the grain yield by 7.1% as compared with MT. Above results indicated the better yield performance of OMST than MST.

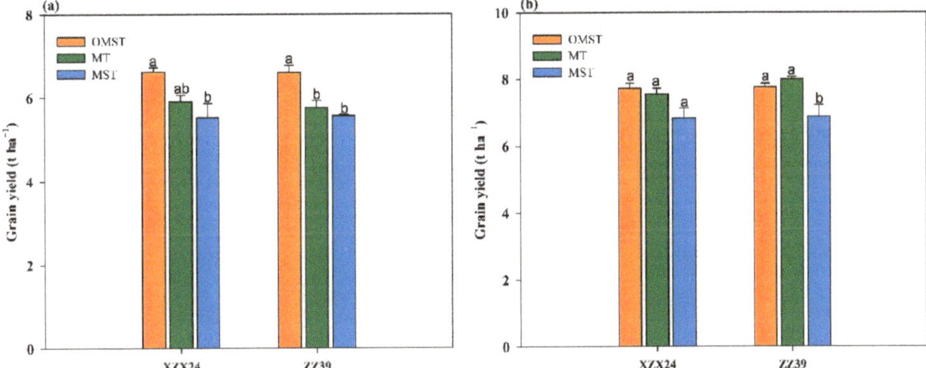

Figure 4. The rice grain yield under different rice establishment methods: (**a**) 2021; and (**b**) 2022. OMST: Orderly mechanical seedling throwing, MT: Manual transplanting, MST: Manual seedling throwing, XZX24: Xiangzaoxian24, ZZ39: Zhongzao39. Different lowercase letters denote statistical differences among treatments of a cultivar at the 5% level according to LSD test. Error bars above mean indicate standard error (*n* = 3).

The grain yield components of the three establishment methods in 2021 and 2022 were shown in Tables 1 and 2, respectively. The panicle number of OMST was significantly higher than that observed in MT and MST (except for the MST of ZZ39 in 2022, which did not reach significant level). For spikelet number per panicle, significant variances between years were observed. The spikelet number per panicle in OMST was significantly lower than MT in 2021 but did not show a significant difference with MT in 2022. Except for XZX24 in 2021, the lowest spikelet number per panicle was observed in MST among the three establishment methods. In addition, although the grain filling percentage did not vary across the three establishment methods in 2021, the grain-filling percentage showed a trend of MT > OMST > MST (*p* < 0.05) in 2022. Besides, the grain weight did not show difference among establishment methods, which was consistent across years. The yield component analysis indicated that the OMST showed an improved panicle formation ability than MT and MST.

Table 1. The grain yield components under different rice establishment methods in 2021.

Variety	Establishment Method	Panicle Number (No. m^{-2})	Spikelet Per Panicle	Grain Filling Percentage (%)	1000-Grain Weight (g)
XZX24	OMST	371.7 a	103.4 b	80.0 a	23.8 a
	MT	340 b	115.8 a	78.1 a	23.6 a
	MST	326.9 b	111.5 a	78.1 a	23.3 a
ZZ39	OMST	285.2 a	110.9 b	85.1 a	25.9 a
	MT	237.1 b	121.1 a	83.9 a	25.7 a
	MST	248.5 b	109.1 b	87.5 a	26.1 a
	T	**	**	ns	ns
	V	**	ns	**	**
	T*V	ns	ns	ns	ns

Note: In each column, different lowercase letters denote statistical differences between treatments of a variety at the 5% level according to LSD test. ** denotes the differences were statistically significant across treatments, ns denotes the differences across treatments did not reach significant level. OMST: Orderly mechanical seedling throwing, MT: Manual transplanting, MST: Manual seedling throwing, XZX24: Xiangzaoxian24, ZZ39: Zhongzao39.

Table 2. The grain yield components under different rice establishment methods in 2022.

Variety	Establishment Method	Panicle Number (No. m^{-2})	Spikelet Per Panicle	Grain Filling Percentage (%)	1000-Grain Weight (g)
XZX24	OMST	360.0 a	127.6 a	79.1 b	23.0 a
	MT	320.0 b	122.1 ab	85.0 a	23.3 a
	MST	323.3 b	120 b	74.7 c	23.1 a
ZZ39	OMST	287.8 a	132.4 a	79.9 b	26.5 a
	MT	274.4 b	130.4 a	81.6 a	26.5 a
	MST	283.3 ab	115.8 b	67.8 c	26.5 a
	T	**	**	**	ns
	V	**	**	**	**
	T*V	**	**	**	ns

Note: In each column, different lowercase letters denote statistical differences between treatments of a variety at the 5% level according to LSD test. ** denotes the differences were statistically significant across treatments, ns denotes the differences across treatments did not reach significant level. OMST: Orderly mechanical seedling throwing, MT: Manual transplanting, MST: Manual seedling throwing, XZX24: Xiangzaoxian24, ZZ39: Zhongzao39.

3.2. Aboveground Biomass

The aboveground biomass of the three establishment methods at middle tillering stage (T), panicle initiation stage (PI), heading stage (HD) and physiological maturity stage (PM) were presented in Figure 5. When compared with MT, the OMST showed a better bio-mass accumulation ability before grain filling, which was significantly increased by 67.4% and 48.5% at T, by 49.8% and 25.3% at PI, and by 4.3% and 12.3% at HD, for XZX24 and ZZ39, respectively, averaged across years. At PM, although the aboveground biomass in OMST was significantly higher than MT in 2021, no significant difference was observed between the two treatments in 2022. Mean-while, the biomass accumulation of MST from T to HD was similar or lower than that observed in OMST, and was similar or higher than that of MT. In maturity, the lowest aboveground biomass was observed in MST (except for ZZ39 in 2021, the biomass in MST did not vary with MT).

The OMST significantly improved the biomass accumulation after heading as compared with MST (Figure 6). When averaged across years, the biomass accumulation after heading of the OMST was significantly increased by 50.6% and 24.2% for XZX24 and ZZ39, respectively, as compared with that of MST. Meanwhile, the biomass accumulation of OMST was significantly by increased 28.6% and 29.5% for XZX24 and ZZ39 in 2021, respectively as compared with MT. In 2022, however, no significant variance was observed between OMST and MT on biomass accumulation after heading.

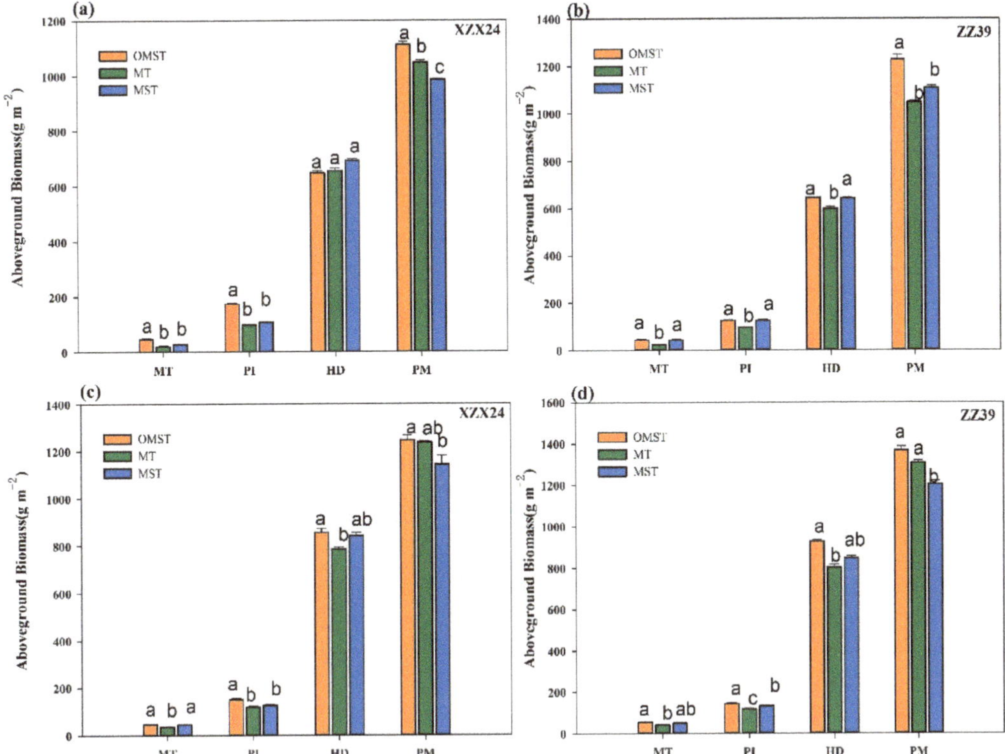

Figure 5. The aboveground biomass at different growth stages under different rice establishment methods: (**a**) XZX24 in 2021, (**b**) ZZ39 in 2021, (**c**) XZX24 in 2022, (**d**) ZZ39 in 2022. OMST: Orderly mechanical seedling throwing, MT: Manual transplanting, MST: Manual seedling throwing, XZX24: Xiangzaoxian24, ZZ39: Zhongzao39, T: Middle tillering stage, PI: Panicle initiation stage, HD: Heading stage, PM: Physiological maturity stage. Different lowercase letters denote statistical differences among treatments of a cultivar at the 5% level according to LSD test. Error bars above mean indicate standard error ($n = 3$).

3.3. The Dynamics of Tiller Number

Significant faster tiller generation speed on OMST than that in MT was observed in both varieties and in both years (Figure 7). At 18 days after transplanting (DAT), the tiller number of OMST was 194.7 per m^2 and 216.7 per m^2 for XZX24 and ZZ39 in 2021, and was 336.6 per m^2 and 246.6 per m^2 for XZX24 and ZZ39 in 2022, which was increased by 21.3–71.0% as compared with that of MT. The advantages of OMST in tiller generation was maintained from 18 DAT to 54 DAT, in which the tiller number per unit area of OMST was increased by 9.6% and 11.2% for XZX24 and ZZ39, respectively, as compared with MT, averaged across years. In MST system, similar or even higher tiller generation speed than that of OMST was observed at the early growth stage. However, when the tiller number reached the maximum level, the tiller number decreased faster in MST than that in OMST. At 54 DAT, 8.8% and 9.5% higher tiller numbers in OMST than in MST was observed for XZX24 and ZZ39, respectively, when averaged across years.

3.4. Leaf Area Index

The leaf area index (LAI) of the three establishment methods at different growth stages are shown in Figure 8. For both varieties, the LAI of OMST was significantly higher than that of MT and did not show significant difference as compared with that of MST at middle tillering stage. At the PI and HD stage, the LAI of the OMST was highest amongst the establishment methods, which was in-creased by 5.4% and 17.6% at PI, respectively, as compared with MT and MST, and was increased by 15.1% and 29.0% at HD, respectively, as compared with MT and MST.

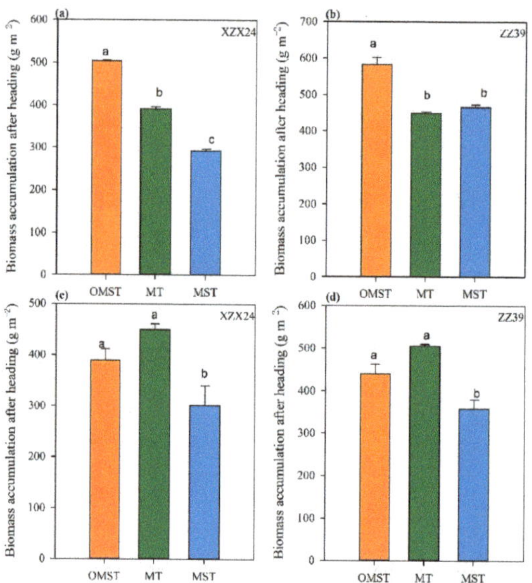

Figure 6. The biomass accumulation after heading under different rice establishment methods: (**a**,**b**) 2021; and (**c**,**d**) 2022. OMST: Orderly mechanical seedling throwing, MT: Manual transplanting, MST: Manual seedling throwing, XZX24: Xiangzaoxian24, ZZ39: Zhongzao39. Different lowercase letters denote statistical differences among treatments of a cultivar at the 5% level according to LSD test. Error bars above mean indicate standard error (*n* = 3).

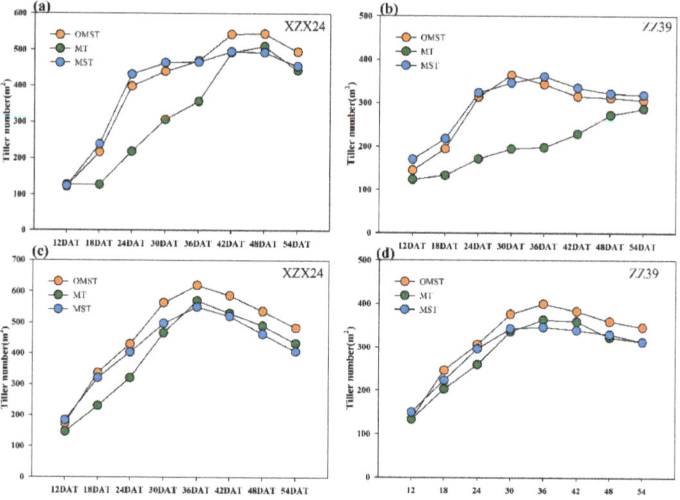

Figure 7. The tillering dynamics of different rice establishment methods: (**a**,**b**) 2021; and (**c**,**d**) 2022. OMST: Orderly mechanical seedling throwing, MT: Manual transplanting, MST: Manual seedling throwing, XZX24: Xiangzaoxian24, ZZ39: Zhongzao39, DAT: Days after transplanting.

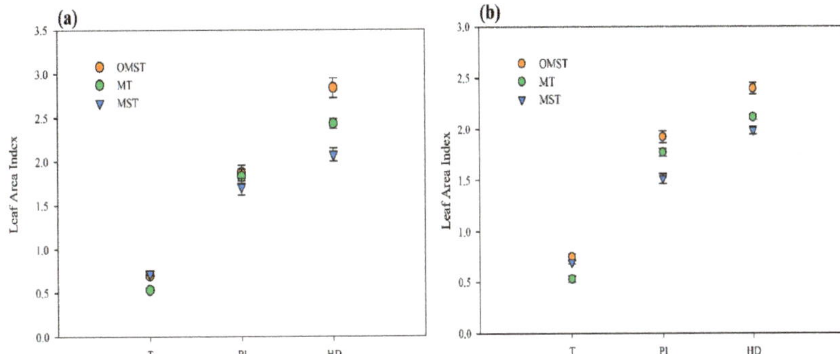

Figure 8. The leaf area index at different growth stages under different rice establishment methods: (**a**) Xiangzaoxian24; and (**b**) Zhongzao39. OMST: Orderly mechanical seedling throwing, MT: Manual transplanting, MST: Manual seedling throwing, XZX24: Xiangzaoxian24, ZZ39: Zhongzao39, T: Middle tillering stage, PI: Panicle initiation stage, HD: Heading stage. Error bars above mean indicate standard error (*n* = 3).

3.5. Leaf SPAD Reading

The SPAD reading of the flag leaf in OMST was lower than that in MT at HD stage (Figure 9). While no significant difference was observed between OMST and MST treatment at HD. At maturity, the SPAD value of OMST did not vary with that of M, and was 16.8% and 5.3% higher than that of MST for XZX24 and ZZ39 respectively (although the variance in ZZ39 did not reach a significant level). The above results suggested that the OMST may have better photosynthetic ability than MST during grain filling stage.

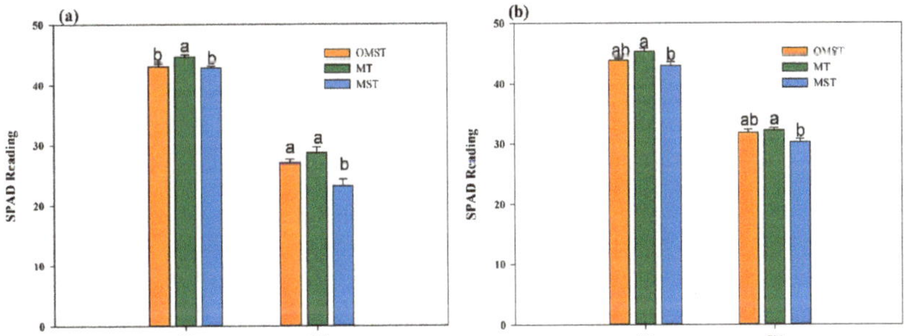

Figure 9. The SPAD reading of the flag leaf under different rice establishment methods: (**a**) Xiangzaoxian24; and (**b**) Zhongzao39. OMST: Orderly mechanical seedling throwing, MT: Manual transplanting, MST: Manual seedling throwing, XZX24: Xiangzaoxian24, ZZ39: Zhongzao39, T: Middle tillering stage, PM: Physiological maturity stage, HD: Heading stage. Different lowercase letters denote statistical differences among treatments of a cultivar at the 5% level according to LSD test. Error bars above mean indicate standard error (*n* = 3).

4. Discussion

Developing an efficient and high-yielding mechanical rice establishment system is one of the most important approaches for intensive and large-scale rice production. In general, rice can be established by either transplanting, direct seeding or seedling throwing [16]. In recent years, several machine transplanting and machine direct-seeding systems have been successfully developed, and the yield and growth performance, the optimum agronomic practices, such as nitrogen rate, and planting density, have also been widely elucidated [9–12]. In rice seedling throwing systems, however, the mechanical equipment and the relative technology is still lacking. The results of the present study showed that the newly developed orderly mechanical seedling throwing system represented significant yield advantages compared to traditional manual seedling throwing, suggesting the potential of OMST system as a replacement to traditional MST for rice production in the future. In present study, significant variances in yield components were observed between OMST and MST. Although the tiller generation speed at early growth stages was similar between OMST and MST, the final tiller number, panicle number and total spikelet number in OMST was significantly higher than in MST, which resulted in the yield advantages in OMST. It has been reported that the uneven distribution of the seedling in MST population inhibit the tiller generation rate and accelerate the death of existed tillers [17]. The higher panicle-bearing tiller rate and sink size in orderly distributed rice population have also been observed as compared with that in in-orderly ones [18].

The shift from the in-orderly population in MST to the orderly ones in OMST is beneficial for the improvement in field micro-environment. It has been suggested that the orderly distribution of rice seedlings in OMST greatly improved the wind velocity as compared with MST [19]. The light transition rate was improved in orderly distributed rice population [20]. Such changes might be responsible for the changes of growth characters in OMST. Our study found that the total biomass and the biomass accumulation after heading was significantly higher in OMST than in MST, which contribute to similar grain filling percentage and grain weight between the two system even though the sink size in OMST was significantly improved. The larger sink size generally requires better dry matter assimilation or translocation ability during grain filling [21], which is influenced by the canopy structure, the light interception rate, as well as the leaf function duration [22–24]. While in present study, the improved dry matter accumulation after heading in OMST than in MST might be due to: (1) the higher LAI that might increase light interception and improve radiation use efficiency; (2) the higher SPAD value at HD and the lower SPAD decrease from HD to PM that might enhance the leaf photosynthetic ability and reduce the leaf senescence rate during grain filling. Previous research found that the root dry matter and root activity was higher in seedling throwing system than in seedling transplanting [25], and the vigor of the root system are highly correlated with the leaf senescence [26]. However, the differences between OMST and MST with regards to root traits, leaf photosynthetic ability, as well as the light interception and radiation use efficiency remained to be unknown, which need to be addressed in future studies.

The grain yield of the OMST showed no significant difference or even higher than that of traditional manual transplanting, which partially reflected the yield stability and the extension potential of this system. The yield component analysis suggested that the OMST possessed better panicle generation ability than MT, which was mainly attributed to more tillers and biomass accumulation especially at vegetative growth stages. This result was consistent with previous research in which the rice seedling throwing system was characterized as earlier tillering, higher plant density and more panicles in comparison with the seedling transplanting system [27,28]. However, the significant higher spikelet number in 2021 and significantly improved grain filling percentage in 2022 in MT system compensated the decrease in panicle number as compared with OMST, which let to equivariant grain yield between MT and OMST. In addition, the grain yield of the MST showed no difference or lower than MT in present study, which was in consistent with several research studies [29–31]. However, the research of Sanbagavalli denoted that the grain yield in MST was higher than that in MT [32]. The variances between different research studies might be attributed to the differences in climate condition, genotype and agronomic practices [27,33] and the yield performance of OMST against MT and MST needs to be examined further. Significant variances in spikelet per panicle between years and treatments were observed. In 2021, The spikelet per panicle in MT was higher or similar than OMST and MST, while the grain-filling percentage showed no difference between treatments, which resulted in increased filled spikelet number per panicle in MT. While in 2022, although the spikelet number per panicle of MT showed no difference to MST and OMST, the significantly improved grain-filling percentage also resulted in improved filled spikelet number per panicle in MT. Nevertheless, the mechanisms underlying the variances in yield components among different establishment methods responding to years and locations remain unknown and need to be addressed in the future. Moreover, the comparison between OMST and other rice establishment methods, such as machine transplanting, direct seeding remain unknown and need to be addressed in future studies.

The growth and the yield formation characters of OMST showed significant variances to that of MT and MST, indicating that the optimum crop management methods for OMST might be varied with the other crop establishment methods. However, the responses of the OMST system to different planting density, fertilizer and water management strategies remains unknown and high-yielding and efficient cultivation technology for OMST need to be developed in the future.

5. Conclusions

The grain yield of the newly developed orderly mechanical seedling throwing system was significantly higher than manual seedling throwing and was equivalent to that of manual transplanting. The yield components analysis showed that the differences in grain yield among the three establishment methods were mainly attributed to the variances in panicle number and total spikelet number. Further analysis suggested that the orderly mechanical seedling throwing takes the advantages of higher biomass accumulation after heading, increased leaf area index and decreased leaf senescence rate against manual seedling throwing, and more tillers and biomass accumulation at vegetative growth stage against manual transplanting. Such advantages might be attributed to the combined effects of seedling throwing and orderly seedling distribution. The present study demonstrated that the orderly mechanical seedling throwing is an efficient and high-yielding rice establishment method that might be a promising option to replace traditional manual seedling throwing in rice production.

Author Contributions: Conceptualization, Q.T.; methodology, W.W. and L.X.; software, W.W.; investigation, L.X.; writing—original draft preparation, W.W.; writing—review and editing, H.Z.; visualization, W.W.; supervision, Q.T. All authors have read and agreed to the published version of the manuscript.

Funding: This research was funded by the Project for China Agriculture Research System (CARS-01-27); the National Natural Science Foundation of China (32201897); the Natural Science Foundation of Hunan Province, China (2021JJ40248).

Institutional Review Board Statement: Not applicable.

Informed Consent Statement: Not applicable.

Data Availability Statement: The data used to support the findings of this study are available from the corresponding author upon request.

Conflicts of Interest: The authors declare no conflict of interest.

References

1. Xu, L.; Zhan, X.; Yu, T.; Nie, L.; Huang, J.; Cui, K.; Wang, F.; Li, Y.; Peng, S. Yield performance of direct-seeded, double-season rice using varieties with short growth durations in central China. *Field Crop. Res.* **2018**, *227*, 49–55. [CrossRef]
2. Peng, S.; Tang, Q.; Zou, Y. Current status and challenges of rice production in China. *Plant Prod. Sci.* **2009**, *12*, 3–8. [CrossRef]
3. Wang, W.; He, A.; Jiang, G.; Sun, H.; Jiang, M.; Man, J.; Nie, L. Ratoon rice technology: A green and resource-efficient way for rice production. *Adv. Agron.* **2020**, *159*, 135–167.
4. Bu, R.; Ren, T.; Lei, M.; Liu, B.; Li, X.; Cong, R.; Lu, J. Tillage and straw-returning practices effect on soil dissolved organic matter, aggregate fraction and bacteria community under rice-rice-rapeseed rotation system. *Agric. Ecosyst. Environ.* **2020**, *287*, 106681. [CrossRef]
5. Huang, M.; Zou, Y. Integrating mechanization with agronomy and breeding to ensure food security in China. *Field Crop. Res.* **2018**, *224*, 22–27. [CrossRef]
6. Xu, Z.; He, P.; Yin, X.; Struik, P.C.; Ding, W.; Liu, K.; Huang, Q. Simultaneously improving yield and nitrogen use efficiency in a double rice cropping system in China. *Eur. J. Agron.* **2022**, *137*, 126513. [CrossRef]

7. Li, H.; Liu, Y.; Zhao, X.; Zhang, L.; Yuan, K. Estimating effects of cooperative membership on farmers' safe production behaviors: Evidence from the rice sector in China. *Environ. Sci. Pollut. Res. Int.* **2021**, *28*, 25400–25418. [CrossRef]

8. Xin, F.; Xiao, X.; Dong, J. Large increases of paddy rice area, gross primary production, and grain production in Northeast China during 2000–2017. *Sci. Total Environ.* **2020**, *711*, 135183. [CrossRef]

9. Liu, H.; Hussain, S.; Zheng, M.; Peng, S.; Huang, J.; Cui, K.; Nie, L. Dry direct-seeded rice as an alternative to transplanted-flooded rice in Central China. *Agron. Sustain. Dev.* **2015**, *35*, 285–294. [CrossRef]

10. Tao, Y.; Chen, Q.; Peng, S.; Wang, W.; Nie, L. Lower global warming potential and higher yield of wet direct-seeded rice in Central China. *Agron. Sustain. Dev.* **2016**, *36*, 24. [CrossRef]

11. Huang, M.; Chen, J.; Cao, F.; Zou, Y. Increased hill density can compensate for yield loss from reduced nitrogen input in machine-transplanted double-cropped rice. *Field Crop. Res.* **2018**, *221*, 333–338. [CrossRef]

12. Ke, J.; Xing, X.; Li, G.; Ding, Y.; Dou, F.; Wang, S.; Chen, L. Effects of different controlled-release nitrogen fertilisers on ammonia volatilisation, nitrogen use efficiency and yield of blanket-seedling machine-transplanted rice. *Field Crop. Res.* **2017**, *205*, 147–156. [CrossRef]

13. Liu, Y.; Li, C.; Fang, B.; Fang, Y.; Chen, K.; Zhang, Y.; Zhang, H. Potential for high yield with increased seedling density and decreased N fertilizer application under seedling-throwing rice cultivation. *Sci. Rep.* **2019**, *9*, 731. [CrossRef]

14. Xiang, L.; Tang, Q.; Wang, W.; Zheng, H.; Zheng, Z. Effects of Orderly Mechanical Seedling-Broadcasting on Disaster Reduction of Double Cropping Late Rice Production under Autumn Low Temperature. *Chin. J. Agrometrorol.* **2022**, *2*, 500–508, (In Chinese with English abstract).

15. Deng, X.; Chen, B.; Chen, Y. Variations in root morphological indices of rice (*Oryza sativa* L.) induced by seedling establishment methods and their relation to arsenic accumulation in plant tissues. *Environ. Pollut.* **2021**, *281*, 116999. [CrossRef]

16. Singh, Y.; Singh, V.P.; Singh, G.; Yadav, D.S.; Sinha, R.K.P.; Johnson, D.E.; Mortimer, A.M. The implications of land preparation, crop establishment method and weed management on rice yield variation in the rice—Wheat system in the Indo-Gangetic plains. *Field Crop. Res.* **2011**, *121*, 64–74. [CrossRef]

17. Tang, S. Seedling broadcasting in China: An overview. In *Direct Seeding: Research Strategies and Opportunities*, 1st ed.; Pandey, S., Mortimer, M., Wade, L., Tuong, T.P., Lopez, K., Hardy, B., Eds.; International Rice Research Institute: Los Baños, Philippines, 2002; pp. 177–185.

18. Ai, Z.; Guo, X.; Liu, W.; Ma, G.; Qing, X. Changes of safe production dates of double-season rice in the middle reaches of the Yangtze River. *Acta Agron. Sin.* **2014**, *40*, 1320–1329. [CrossRef]

19. Wang, W.; Tang, Q.; Chen, Y.; Jia, W.; Luo, Y.; Wang, X.; Zheng, H.; Xiong, J. Evaluation of orderly mechanical seedling-broadcasting on yield formation and growth characteristics of rice. *Acta Agron. Sin.* **2021**, *47*, 942–951.

20. Chen, D.; Yang, W.; Ren, W. Effects of rice seedlings horizontal distribution on the dynamics of rice population, canopy light transmittance rate and panicle characteristics. *Chin. J. Appl. Ecol.* **2017**, *18*, 359–365.

21. Yao, F.; Huang, J.; Nie, L.; Cui, K.; Peng, S.; Wang, F. Dry matter and N contributions to the formation of sink size in early- and late-maturing rice under various N rates in central China. *Int. J. Agric. Biol.* **2016**, *18*, 46–51. [CrossRef]

22. Gautam, P.; Lal, B.; Nayak, A.K.; Raja, R.; Panda, B.B.; Tripathi, R.; Shahid, M.; Kumar, U.; Baig, M.J.; Swai, C.K.; et al. Inter-relationship between intercepted radiation and rice yield influenced by transplanting time, method, and variety. *Int. J. Biometeorol.* **2019**, *63*, 337–349. [CrossRef]

23. Acevedo-Siaca, L.G.; Dionora, J.; Laza, R.; Paul, Q.W.; Long, S.P. Dynamics of photosynthetic induction and relaxation within the canopy of rice and two wild relatives. *Food Energy Secur.* **2021**, *10*, e286. [CrossRef] [PubMed]

24. Lee, S.; Masclaux-Daubresse, C. Current Understanding of Leaf Senescence in Rice. *Int. J. Mol. Sci.* **2021**, *22*, 4515. [CrossRef] [PubMed]

25. Luo, Y.; Wang, W.; Zheng, H.; Liu, G.; Chao, Y.; Xu, C.; Zheng, Z.; Li, X.; Wei, Y.; Tang, Q. Influences of Different Mechanical and Orderly Planting Methods on Growth Characteristics and Yield of rice. *J. Agric. Sci. Technol.* **2021**, *23*, 162–171.

26. Liu, H.; He, A.; Jiang, G.; Hussain, S.; Wang, W.; Sun, H.; Nie, L. Faster leaf senescence after flowering in wet direct-seeded rice was mainly regulated by decrease in cytokinin content as compared with transplanted-flooded rice. *Food Energy Secur.* **2020**, *9*, e232. [CrossRef]

27. Horgan, F.G.; Figueroa, J.Y.; Almazan, M.L.P. Seedling broadcasting as a potential method to reduce apple snail damage to rice. *Crop Prot.* **2014**, *64*, 168–176. [CrossRef]

28. Vishwakarma, A.; Singh, J.K.; Sen, A.; Bohra, J.S.; Singh, S. Effect of transplanting date and age of seedlings on growth, yield and quality of hybrids under system of rice (*Oryza sativa*) intensification and their effect on soil fertility. *Indian J. Agric. Sci.* **2016**, *86*, 679–685.

29. Chen, S.; Ge, Q.; Chu, G.; Xu, C.; Yan, J.; Zhang, X.; Wang, D. Seasonal differences in the rice grain yield and nitrogen use efficiency response to seedling establishment methods in the Middle and Lower reaches of the Yangtze River in China. *Field Crop. Res.* **2017**, *205*, 157–169. [CrossRef]

30. Chauhan, B.S.; Abeysekera, A.S.K.; Wickramarathe, M.S.; Kulatunga, S.D.; Wickrama, U.B. Effect of rice establishment methods on weedy rice (*Oryza sativa* L.) infestation and grain yield of cultivated rice (*O. sativa* L.) in Sri Lanka. *Crop Prot.* **2014**, *55*, 42–49. [CrossRef]

31. Huang, M.; Zhou, X.F.; Cao, F.B.; Xia, B.; Zou, Y.B. No-tillage effect on rice yield in China: A meta-analysis. *Field Crop Res.* **2015**, *183*, 126–137. [CrossRef]

32. Sanbagavalli, S.; Kandasamy, O.S. Nitrogen management and economic returns of seedling throwing method of rice planting in dry season. *Agric. Sci. Dig.* **2000**, *20*, 42–45.
33. Senguttuvel, P.; Sravanraju, N.; Jaldhani, V.; Divya, B.; Beulah, P.; Nagaraju, P.; Manasa, Y.; Hari Prasad, A.S.; Brajendra, P.; Subrahmanyam, D.; et al. Evaluation of genotype by environment interaction and adaptability in low land irrigated rice hybrids for grain yield under high temperature. *Sci. Rep.* **2021**, *11*, 15825. [CrossRef] [PubMed]

Article

Identification of a New Wide-Compatibility Locus in Inter-Subspecific Hybrids of Rice (*Oryza sativa* L.)

Weibo Zhao [1,2], Wei Zhou [1,2], Han Geng [1,2], Jinmei Fu [1,2], Zhiwu Dan [1,2], Yafei Zeng [1,2], Wuwu Xu [1,2], Zhongli Hu [1,2] and Wenchao Huang [1,2,*]

[1] State Key Laboratory of Hybrid Rice, Wuhan University, Wuhan 430072, China
[2] College of Life Sciences, Wuhan University, Wuhan 430072, China
* Correspondence: wenchaoh@whu.edu.cn

Abstract: As a special class of rice germplasm, wide-compatibility varieties (WCVs) guarantee the fertility of hybrids when there is cross-fertilization between two subspecies. In this study, Chenghui9348 was identified as a new member of the WCV family that improves pollen fertility in an inter-subspecific hybrid. Cytological analysis showed that the abnormal mitosis of microspores resulted in the sterility of pollens at the early bicellular stage in the inter-subspecific hybrid. Furthermore, the new *F12* locus, corresponding to improvements in fertility of the *indica-japonica* hybrid, was found to co-segregate with the RM1047 marker and associated with a region of approximately 630 kb flanked by the D1101 and D1164 markers on chromosome 12. In this region, two putative genes were predicted as the candidates for wide-compatibility genes (WCGs). Sequence analysis revealed that, compared with *indica/japonica* alleles, deletion/insertion occurred within exons of both putative genes. Together, the present study identified another new WC locus, *F12*, and offers more opportunities for further exploitation of inter-subspecific hybrids in rice.

Keywords: rice; hybrid sterility; wide compatibility

Citation: Zhao, W.; Zhou, W.; Geng, H.; Fu, J.; Dan, Z.; Zeng, Y.; Xu, W.; Hu, Z.; Huang, W. Identification of a New Wide-Compatibility Locus in Inter-Subspecific Hybrids of Rice (*Oryza sativa* L.). *Agronomy* **2022**, *12*, 2851. https://doi.org/10.3390/agronomy12112851

Academic Editor: Yong-Bao Pan

Received: 5 October 2022
Accepted: 9 November 2022
Published: 15 November 2022

Publisher's Note: MDPI stays neutral with regard to jurisdictional claims in published maps and institutional affiliations.

1. Introduction

Hybrids between subspecies often lead to hybrid incompatibilities, such as sterility and inviability. Hybrid incompatibility hinders the gene exchange between subspecies, which is a major obstacle to heterosis utilization [1]. Cultivated rice is divided into two species, African rice (*Oryza glaberrima* Steud) and Asian rice (*Oryza sativa* L.). Inter-specific hybrids were made for African rice; the New Rice for Africa (NERICA) varieties exhibited high performance in terms of abiotic and biotic stress resistance, and achieved higher yields compared with its parental varieties [2]. The Asian cultivated rice (*Oryza sativa* L.) was classified into two main subspecies, *indica* and *japonica* [3]. The inter-subspecific hybrids have stronger hybrid vigor than the hybrid within the subspecies; however, the sterility barriers in inter-subspecific hybrids prevent the application of heterosis to hybrid rice breeding programs, and the utilization of rice heterosis was limited to *indica-indica* hybrids, which increased the grain yield by about 20% [1,3]. Fortunately, the discovery of wide-compatibility varieties (WCVs), which allow for the progeny of *indica* and *japonica* crossings to exhibit the fertility of normal hybrids, brought breeders powerful tools for exploiting the heterosis between the two subspecies and for further improving grain yield [4].

Nowadays, several genetic mechanisms, such as the duplicate gametophytic lethal model, allelic interaction at a single locus, and epistatic interaction between loci, have been proposed to explain *indica/japonica* hybrid sterility [5,6].

In 1962, Kitamura first put forward a one-locus sporo-gametophytic interaction model, which could explain the genetic behavior of most hybrid sterility loci [5], and then Ikehashi and Araki demonstrated that a locus named *S5* was consistent with this model [4]. According to this model, there are three alleles at the *S5* locus: an *indica* allele, S_5^i, a *japonica* allele,

S_5^j, and a neutral allele (wide-compatibility allele), S_5^n. The hybrid progeny between *indica* and *japonica* would be partially sterile, with a S_5^i/S_5^j genotype. However, the progenies bearing an S_5^i or S_5^j allele would be fully fertile when WCVs (bearing at least one S_5^n allele) are crossed to *indica* or *japonica*. Ikehashi and Araki also found that the *S5* locus was located on chromosome 6 by using morphological markers [4], and this result has been further confirmed by using isozymes and molecular markers [7,8]. Then, Qiu and Ji analyzed the *S5* locus by using near-isogenic lines (NILs) to delimitate the *S5* locus to a 40 kb- and 50 kb-genomic DNA segment, respectively [9,10]. Three open reading frames (ORF3, ORF4, and ORF5) comprise the triallelic system of the *S5* locus as shown by Chen et al., and Yang et al. proposed to refer to them as a "Killer-Protector System" [11,12].

However, the progeny showed low fertility when some WCVs were crossed to different *indica* or *japonica* varieties [13]. Thus, hybrid sterility was considered to be caused by allelic interactions at many different loci [14–16]. By using different cross combinations involving WCVs, many sterility gene loci other than *S5* have been found. Since then, more than twenty loci conferring embryo sac sterility and thirty loci conferring male sterility have been found to influence hybrid sterility [17]. Hopefully, it will be possible to overcome the hybrid sterility in rice breeding when there is an understanding of the mechanism of these sterility gene loci. However, we still cannot solve the sterility of the inter-subspecific hybrids, even though the main wide-compatibility allele, i.e., S_5^n, has been cloned, and it is necessary to exploit more WCGs to explain the underlying mechanism. Here, we report a newly identified locus conferring intercrossing compatibility in Chenghui9348, a rice variety carrying an *S5-i* allele that uses 9311 and Lemont as parental varieties. We found that this locus, named *F12-C*, has extensive effects on both pollen and spikelet fertility. These results will help clone this gene and aid in its marker-assisted selection in rice-breeding programs.

2. Materials and Methods

2.1. Plant Materials and Mapping Populations

The 9311 is an *indica* cultivar that has an *S5-i* allele. Nipponbare and Balilla are typical *japonica* cultivars carrying the *S5-j* allele at the *S5* locus. Chenghui9348, which was developed by the Sichuan Academy of Agricultural Sciences, also has an *S5-i* allele, as shown by sequencing the *S5* locus. A series of hybrids between parents were made in the winter of 2013 in Hainan and planted in the rice-growing season of 2014 in Ezhou, Hubei province. A population of 372 plants from the three-way cross, 9311/Chenghui9348//Nipponbare, were planted in an experimental field of the Hybrid Rice Hainan Experimental Base of Wuhan University in Lingshui, Hainan Province, in November 2013. Another two three-way cross populations (9311/Chenghui9348//Nipponbare, 9311/Chenghui9348//Balilla) were planted in the summer rice-growing season of 2014 in the experimental field of the Hybrid Rice Ezhou Experimental Base of Wuhan University in Ezhou, Hubei Province. The order of the parents in these two three-way cross populations ensures that the population has an S_5^i/S_5^j genotype so that it can counteract the effect of the *S5* site. All materials were planted with an interval of 16.5 cm. Plots were spaced 23.5 cm apart. The crop was managed following normal commercial practices.

2.2. Evaluation of Pollen and Spikelet Fertility

Six florets from three panicles per plant were collected 1–2 days before flowering and fixed in 70% ethanol. Anthers were mixed and stained with 1% iodine potassium iodide (I_2-KI) solution, and more than 300 pollen grains from each individual were observed by light microscope to estimate the percentage of fertile grains. The fully stained pollens were fertile, and partially stained ones were abortive. The rice spikelet fertility is the ratio of fertile spikelets to total spikelets obtained by counting three panicles on the upper half of the panicles for each plant, as described by Wan [15].

2.3. Cytological and Histological Analysis

The F1 (9311/Nipponbare, Chenghui9348/Nipponbare) hybrid spikelets of different stages were collected from florets and then fixed in formalin fixative (formalin: ethanol (50%, *v/v*): glacial acetic acid = 18:1:1, *v/v/v*). The pollen grains were stained in acetocarmine. Analysis of pollen germination on the stigma was performed and the pollen was stained with aniline blue 1 h post-anthesis; ten florets were accounted for per plant and observed by confocal laser scanning microscopy, as described by Zhou [18].

2.4. DNA Preparation and PCR Analysis

DNA was extracted from fresh leaves of each plant and dissolved in TE buffer (10 mM Tris, 0.1 mM EDTA) before the quality test. The eligible samples were diluted to 20 ng/μL with double distilled water (ddH$_2$O) and stored at 4 °C. Polymerase chain reaction (PCR) was performed in 10 μL reaction volumes containing 10 mM Tris-HCl (pH 8.3), 1.5 mM MgCl$_2$, 50 mM KCl, 50 μM dNTP, 0.2 μM SSR (simple sequence repeat) primers, 0.5 U Taq polymerase (TaKaRa, Dalian, China), and 20 ng template. The procedure of amplification consisted of a denaturation step (94 °C, 5 min), followed by 35 cycles of 30 s at 94 °C, 30 s at 55 °C, and 30 s at 72 °C, with a final extension of 10 min at 72 °C. The PCR products were electrophoresed on 6% polyacrylamide denaturing gels in 0.5 × TBE buffer and visualized with silver staining. Amplified DNA fragments showing clear polymorphisms between the parental types were used for the analysis of the three-way cross population and construction of the linkage mapping.

2.5. Molecular Marker Development and Assay

All the SSR markers around the *F12-C* locus region were obtained from Gramene (http://www.gramene.org/microsat/ (accessed on 1 January 2014)) based on the SSR linkage map constructed by McCouch [19]. To obtain more markers at the *F12-C* locus region, we re-sequenced one of the parental Chenghui9348 individuals and compared it with the sequence of 9311 to find the insertion/deletion (InDel) markers. Divergent InDel markers were designed based on the sequence of 9311 by using the Primer 5.0 software.

2.6. BSA and Linkage Analysis

The individuals with the ten highest (H) and ten lowest (L) pollen and spikelet fertilities were selected from the three-way cross population in 2013 to perform bulked segregation analysis (BSA) [7]. A total of 151 SSR markers that generated polymorphic bands between 9311 and Chenghui9348 were used to screen for polymorphisms between the H and L bulks. Polymorphic SSR markers that were confirmed in the bulks were used to analyze the individuals with the lowest (<30%) and highest (>70%) pollen and spikelet fertilities. Individuals carrying the '9311' allele were scored as 0 and those carrying the 'Chenghui9348' allele were scored as 2. QTLs controlling hybrid fertility were determined by interval mapping using QTL IciMapping 4.0 at an LOD threshold of 3.0.

2.7. Sequence Analysis of the Candidate Genes

The coding sequences of candidate genes were amplified in 38 diverse accessions of rice (*O. sativa*) (Supplemental Table S4). Then, the PCR products were sequenced in TSINGKE (http://www.tsingke.net/ (accessed on 1 January 2014)). The primers used for amplifying the target fragments were designed based on the *japonica* cultivar Nipponbare genomic sequence data that were available on Gramene (Supplemental Table S2). The sequenced results were aligned to the parents to find insertion/deletion and SNP sites by using AlignX software.

2.8. RNA Extraction and RT-PCR

Total RNA from the mature root, mature stem, mature leaf, and young panicles was isolated using the TRIzol kit. The extracted RNA was treated with DNase (Thermo Fisher Scientific, Waltham, MA, USA) to eliminate genomic DNA contamination. Reverse transcription was performed using oligo (dT) 18 primers (Thermo Fisher Scientific, Waltham, MA, USA) and reverse transcriptase (Invitrogen Life Technologies, Thermo Fisher Scientific, Waltham, MA, USA) *Actin* primers were as listed (Supplemental Table S2). Amplification of the *actin* gene occurred at 95 °C for 5 min, 95 °C for 35 s, 55 °C for 35 s, and 72 °C for 35 s for 28 cycles. The 310RT primer was used to amplify ORF1 cDNA fragments. RT-PCR was performed at 95 °C for 5 min, 95 °C for 35 s, 56 °C for 35 s, and 72 °C for 35 s for 28 cycles.

3. Results

3.1. Chenghui9348 Has Wide-Compatibility Genes (WCGs)

Pollen and spikelet fertility were normal in 9311, Chenghui9348, and Nipponbare (over 88%) (Table 1). However, the pollen and spikelets of F_1 from the 9311/Nipponbare cross showed low fertility rates, with scores of 24.5 ± 6.2% and 34.5 ± 3.7%, respectively (Table 1). When Chenghui9348 was crossed with Nipponbare, the pollen and spikelet fertilities reached 73.8 ± 4.1% and 70.5 ± 4.2%, respectively (Figure 1, Table 1). Further study showed that pollen and spikelet fertilities were 149 91.0 ± 1.8% and 87.4 ± 1.8% in the Chenghui9348/9311 hybrids (Table 1). These results suggested that Chenghui9348 is a wide-compatibility variety (WCV). Moreover, we found that the crosses of Chenghui9348/*japonica* produced a higher level of fertile hybrids compared to those of the control cross, '9311/*japonica*', indicating that Chenghui9348 has a wide spectrum of compatibility (Supplemental Table S1). To confirm whether it was the S_5^n wide-compatibility gene (WCG) that caused this phenomenon, we sequenced this gene in Chenghui9348 and the results showed that it had a S_5^i genotype (data not shown). So, we speculated that there were new WCGs in the Chenghui9348 cultivar. Further reciprocal cross results indicated that the pollen and spikelet fertilities of the F_1 plants from the 9311/Nipponbare and Chenghui9348/Nipponbare crosses were close to that of their reciprocal F_1, suggesting that the wide spectrum of compatibility was controlled by the nuclear genome (Table 1).

Table 1. Pollen and spikelet fertilities of parents and F_1 hybrids.

Parental Varieties and Crosses	Pollen Fertility (Mean (%) ± SD)	Spikelet Fertility (Mean (%) ± SD)
9311	92.3 ± 2.2	92.0 ± 1.9
Chenghui9348	93.5 ± 2.4	90.2 ± 2.6
Nipponbare	88.0 ± 1.8	91.2 ± 3.2
Balilla	91.0 ± 2.9	90.6 ± 3.8
Chenghui9348/9311	91.0 ± 1.8	87.4 ± 1.8
9311/Nipponbare	24.5 ± 6.2 [†]	34.5 ± 3.7 [†]
Nipponbare/9311	20.5 ± 5.4 [†]	28.3 ± 4.8 [†]
Chenghui9348/Nipponbare	73.8 ± 4.1 [†]	70.5 ± 4.2 [†]
Nipponbare/Chenghui9348	72.3 ± 2.5 [†]	71.9 ± 4.6 [†]

[†] No significant difference between reciprocal cross at 1% level (*t*-test).

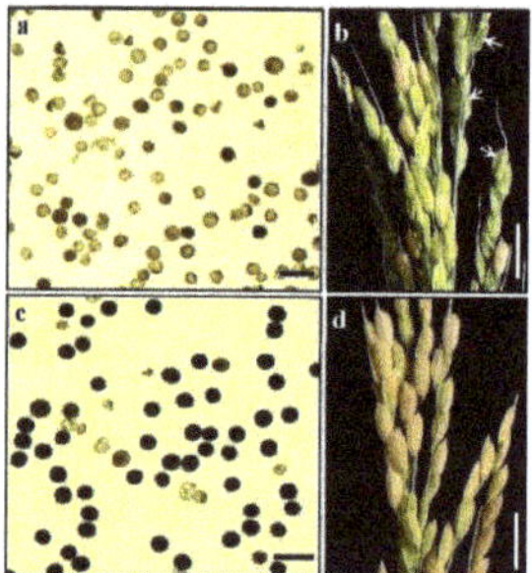

Figure 1. Characterization of the pollen and spikelet fertilities of the inter-subspecific hybrids. (**a,b**) The pollen and spikelet fertilities of the 9311/Nipponbare F_1. White arrows indicate the blighted part of the grains. (**c,d**) The pollen and spikelet fertilities of the Chenghui9348/Nipponbare F_1. (**a,c**) The sterile pollen grains were empty and light-colored. (**a,c**) Scale bar = 50 µm. (**b,d**) Scale bar = 1 cm.

3.2. WCGs in Chenghui9348 Affect Pollen Development and Pollen Adherence to Stigma

We investigated the developmental process of the pollen from the *indica/japonica* hybrid by using the carmine acetate dyeing method. The development of the microspores from the Chenghui9348 × Nipponbare F_1 hybrid was normal, as was that of the 9311 × Nipponbare F_1 hybrid, from the microspore mother cell (MMC) formation stage to the microspore stage (Figure 2a–f). We found that the development of the Chenghui9348/Nipponbare F_1 microspores remained normal; however, most of the 9311 × Nipponbare F_1 hybrid microspores only retained a vegetative nucleus and rarely contained a reproductive one at the early bicellular pollen stage, indicating that there were developmental disorders during the first mitosis (Figure 2g,h). Moreover, with the development of the microspores, most microspores of the 9311 × Nipponbare F_1 hybrid had become empty and shapeless shells at the late bicellular pollen stage. Additionally, compared with the F_1 of the Chenghui9348/Nipponbare hybrids, the contents of the microspores, such as starch, almost disappeared in the microspores of the 9311 × Nipponbare F_1 hybrid, and spherical abortion pollen grains can be clearly observed at the mature pollen stage (Figure 2i,j). The staining results indicated that the pollen grains of the Chenghui9348 × Nipponbare F_1 hybrid developed normally, and most pollen grains of the 9311 × Nipponbare F_1 hybrid were aborted at the early bicellular pollen stage.

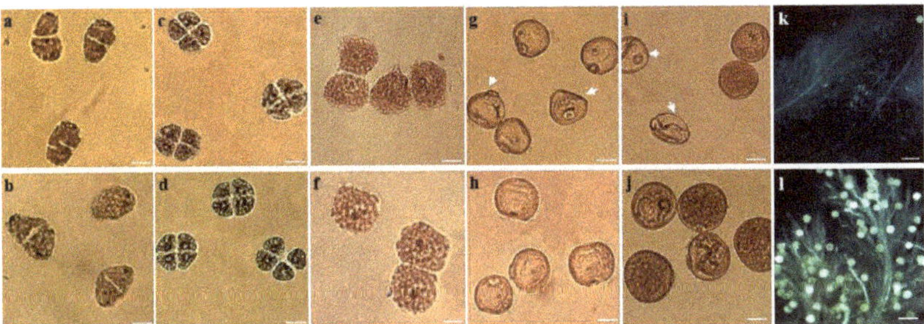

Figure 2. Phenotype of the pollen in the 9311/Nipponbare F_1 hybrid and the Chenghui9348/Nipponbare F_1 hybrid. (**a,c,e,g,i**) Pollen development of the 9311/Nipponbare F_1 hybrid. (**b,d,f,h,j**) Pollen development of the Chenghui9348/Nipponbare F_1 hybrid. (**a,b**), Dyads. (**c,d**), Tetrads. (**e,f**), Early microspores. (**g,h**), Late binucleate pollen. (**i,j**), Mature pollen (arrowheads indicate aborted pollen grains). Scale bar = 50 μm. (**k**) Comparison of pollen grain number on the stigma between the F_1 of the 9311/Nipponbare and Chenghui9348/Nipponbare hybrids (**l**). White arrows show sterility spikelets. Scale bar = 50 μm.

We further monitored the germination of the pollen grains on the stigmas of the inter-subspecific hybrids by aniline blue staining and observed the grains with a fluorescence microscope. The results showed that more pollen grains adhered to the stigmas and were able to germinate in the F_1 of Chenghui9348 × Nipponbare hybrid compared with the F_1 of 9311 × Nipponbare hybrid (Figure 2k,l). Thus, the reduced spikelet fertility of F_1 hybrids from the cross of 9311 and Nipponbare resulted from pollen abortion and poor pollen adherence to the stigma.

3.3. Identification of a New Wide-Compatibility Locus, F12-C, in Chenghui9348

The abnormally low temperature in the winter of 2013 led to the low pollen and spikelet fertilities, and their frequency distributions showed a continuous distribution from 10 to 50%, with peaks around 20% (Figure 3a,b). In 2014, the distribution of pollen fertility in the 9311/Chenghui9348//Nipponbare population was bimodal, ranging from 10% to 100% with a valley at approximately 60% fertility (Figure 3c). Another three-way cross population (9311/Chenghui9348//Balilla) showed almost the same result (Figure 3d). The segregation of highly fertile and partly sterile individuals deviated from the expected 1:1 ratio, with more individuals in the high-fertility group (from 70–100%) than in the low-fertility group (from 10–50%) (Table 2). This result implied that the deviation is likely due to partial abortion of male gametes of 9311-type in the F_1 plant, which increased the ratio of the Chenghui9348/Nipponbare genotypes against the 9311/Nipponbare genotypes.

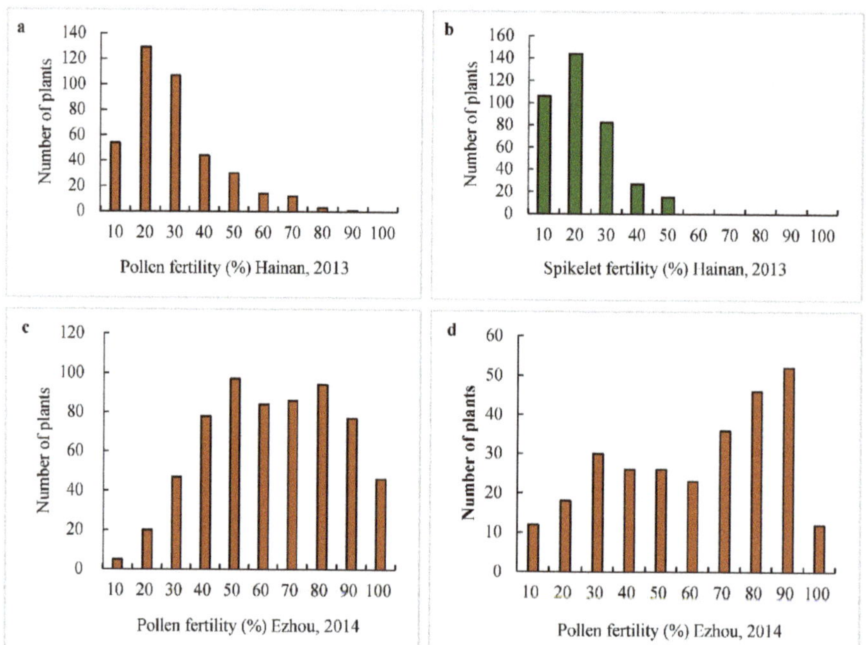

Figure 3. Distribution of pollen and spikelet fertilities in the three-way cross population. (**a,b**) The 9311/Chenghui9348//Nipponbare population (Hainan, 2013). (**c**) The 9311/Chenghui9348//Nipponbare population (Ezhou, 2014). (**d**) The 9311/Chenghui9348//Balilla population (Ezhou, 2014).

Table 2. *Chi*-square test results of low and high pollen fertilities in different populations.

Populations (2014)	Pollen Fertility (10–50%)	Pollen Fertility (70–100%)	χ^2	*p*
9311/Chenghui9348//Nipponbare	247	303	5.70	0.017 *
9311/Chenghui9348//Balilla	112	146	4.48	0.034 *

* Denotes difference at 5% level (*t*-test).

To identify the wide-compatibility loci in Chenghui9348, a total of 151 SSR markers distributed on 12 chromosomes that generated polymorphic bands between 9311 and Chenghui9348 were applied to identify markers linked to the loci by the BSA method. Fourteen SSR markers showed polymorphisms between the bulks. Then, we genotyped the individuals with the 50 lowest (<30%) and 20 highest (>70%) pollen and spikelet fertilities by using these polymorphism markers to detect linkage between markers of genetic compatibility. Two polymorphic SSR markers, RM511 and RM313, located on the same chromosome, chromosome 12, were identified to be associated with the *F12-C* wide-compatibility locus (Fertility 12-Chenghui9348). To confirm the reliability of the linkage between these two markers and *F12-C*, we screened the whole population and found it was significantly associated with *F12-C* by a *t*-test at $\alpha = 0.01$ in the population. Then, with the average distributions of additional SSR markers on chromosome 12, we established a regional linkage map and conducted composite interval mapping with QTL IciMapping 4.0. A QTL was located in a 2.5-cm interval between markers RM1246 and RM7102 on the long arm of chromosome 12

(Figure 4). This QTL was linked to the SSR marker RM1047 with LOD scores of 7.8 and 7.9, explaining 17.9% and 18.4% PVE (phenotypic variance explained) of the variation in pollen and spikelet fertilities, respectively. We also used two three-way cross populations in 2014, as described above, to confirm the location of this QTL.

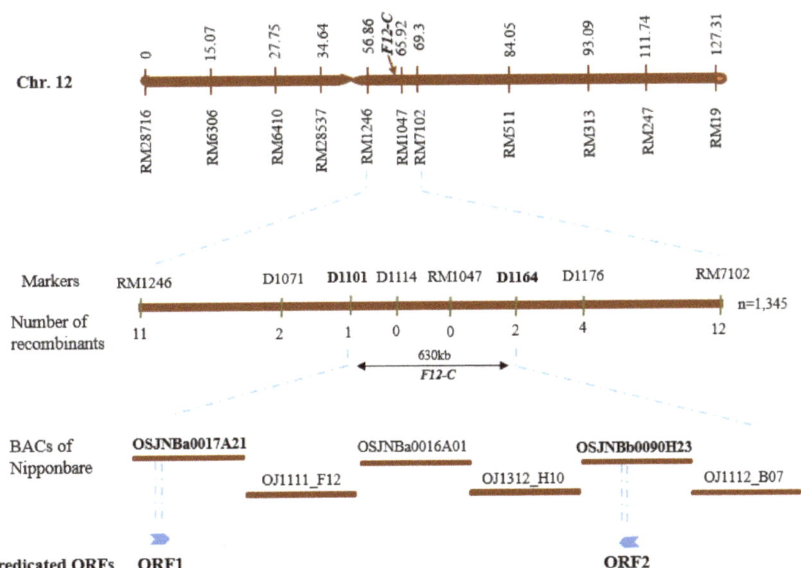

Figure 4. The location of *F12-C* and candidate genomic region. The numbers between markers indicate the recombination events detected between the *F12-C* locus and the respective markers. The short vertical line represents BAC clones of Nipponbare with the accession numbers indicated.

For this locus, the 'Chenghui9348' genotype exhibited higher fertility than the 9311 genotype. In 2013, the average pollen and spikelet fertilities of Chenghui9348-genotype plants at RM1047 were 30.7% and 21.2%, respectively, which are significantly higher than those of the 9311 genotype (17.6% and 13.1%, respectively). In 2014, the average pollen fertility of Chenghui9348-genotype plants was 69%, also significantly higher than that of the 9311 genotype (45%). All the differences were highly significant as determined by the *t*-test (Table 3).

Table 3. Pollen and spikelet fertilities averaged for the *F12-C* locus.

RM1047	Pollen Fertility (%) (2013)	Spikelet Fertility (%) (2013)	Pollen Fertility (%) (2014, Nipponbare)	Pollen Fertility (%) (2014, Balilla)
0 [a]	17.6 **	13.1 **	45 **	44 **
2 [a]	30.7 **	21.2 **	69 **	62 **

[a] Genotype 0 denotes an allele from 9311. Genotype 2 denotes an allele from Chenghui9348. ** Denotes a significant difference between two genotypes at 1% level (*t*-test).

3.4. Mapping the F12-C Locus to a 630 kb Interval

To narrow down the genomic region of *F12-C*, we chose 481 hybrid plants from two three-way cross populations with extreme phenotypes (pollen fertility lower than 30% or higher than 70%) for further mapping. Sixteen recombinant individuals were found by using RM1246 and RM7102 to genotype the extreme individuals. To obtain more polymorphic markers between Chenghui9348 and 9311 in the region of RM1246-RM7102, we compared the sequence of Chenghui9348 with that of 9311 in this region to seek the InDel markers, and fifteen divergent InDel markers were chosen for further mapping (Supplemental Table S2). Thus, eighteen markers, including three SSR markers, were available to analyze the recombination individuals. The analysis revealed one recombinant event between D1101 and *F12-C*, and two recombinant events between D1164 and *F12-C*. In addition, there are six markers that co-segregated with *F12-C*, including RM1047 (Supplemental Table S2). Thus, we narrowed down the genomic region containing the *F12-C* locus to a region approximately 630 kb in length by InDel markers D1101 and D1164 (Table 4). NCBI Map Viewer (http://www.ncbi.nlm.nih.gov/mapview/, accessed on 14 November 2013) showed that there are six BAC clones in this region, according to the Nipponbare genome (OSJNBa0017A21, OJ1111_F12, OSJNBa0016A01, OJ1312_H10, OSJNBb0090H23, OJ1112_B07) (Figure 4).

Table 4. Molecular marker genotypes of partial recombinants.

| Markers | Sterile Individuals | Fertile Individuals | | | | | |
	n30-02	n79-09	n36-05	n27-04	n78-04	n60-09	n79-07
RM1246	0	2	2	2	2	2	0
D1047	0	2	2	2	2	2	0
D1071	0	2	2	2	2	2	0
D1081	0	2	2	2	2	2	0
D1101	0	2	2	2	2	2	0
D1114	0	2	2	2	2	2	2
RM1047	0	2	2	2	2	2	2
D1150	0	2	2	2	2	2	2
D1164	2	0	2	2	2	2	2
D1176	2	0	0	0	2	2	2
D1265	2	0	0	0	0	0	2
RM7102	2	0	0	0	0	0	2

Genotype 0 denotes an allele from 9311. Genotype 2 denotes an allele from Chenghui9348.

3.5. Candidate Genes in the Location Region

Gene prediction analysis of the 630-kb region from Nipponbare using the Rice Genome Annotation Project (RGAP, http://rice.plantbiology.msu.edu/, accessed on 6 February 2013) showed that there were 29 predicted ORFs in these six BAC clones (Supplemental Table S3). Because *F12-C* has a similar function to *Sn 5*, which has a 136-bp deletion compared with *Si 5* and *Sj 5*, we sequenced the predicted ORFs in this region to find the target gene by comparing the genomic sequences of Chenghui9348 and Nipponbare. Finally, two ORFs, ORF1 and ORF2, were identified. ORF1 is a gene with a transcript of 1.04 kb with three exons encoding a 232-amino-acid hypothetical protein, and ORF2 has five exons with a transcript length of 1344 bp encoding a 447-amino-acid hypothetical protein. Using the genomic sequence of Nipponbare as the reference sequence, we found that the coding sequence of ORF1 in Chenghui9348 has 16 SNPs and a deletion of 46 bp in the third exon, which resulted in a frameshift mutation (Figure 5a). ORF2, however, has 44 SNPs and an insertion of 12 bp in the first exon (Figure 5b).

3.6. Allelic Sequencing of F12-C Candidate Genes

We investigated the diversity of the coding sequences of ORF1 and ORF2 among 38 rice cultivars, including 16 *indica* cultivars and 22 *japonica* cultivars, to deduce their sequence features (Supplemental Table S4). The results suggested that ORF1 and ORF2 can be classified into three and four types according to their sequences, respectively. Specifically, the re-sequencing results of ORF1 showed that all cultivars who had the same genotype as Chenghui9348 just have one genotype: Type3 (Figure 5a). In contrast, the ORF2 alleles that had a 12-bp insertion similar to Chenghui9348 had two genotypes, Type2 and Type3, indicating the diversity of Chenghui9348 alleles in ORF2 (Figure 5b).

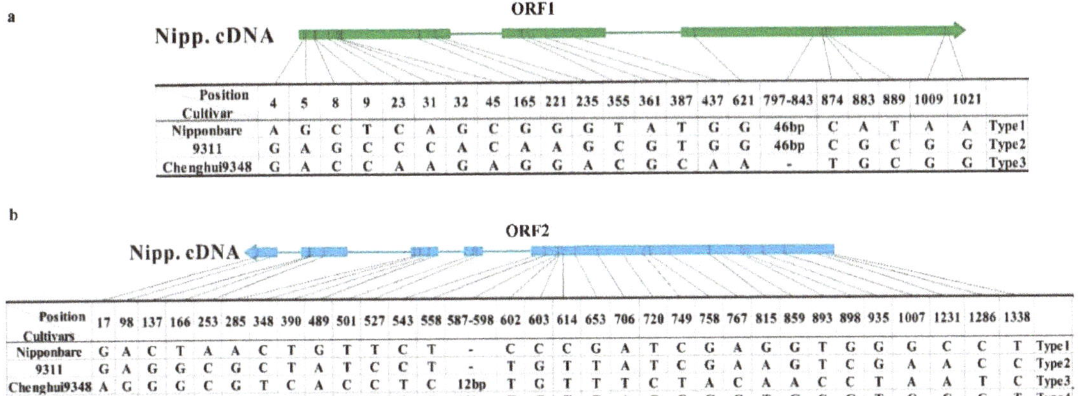

ORF1

Position Cultivar	4	5	8	9	23	31	32	45	165	221	235	355	361	387	437	621	797-843	874	883	889	1009	1021	
Nipponbare	A	G	C	T	C	A	G	C	G	G	G	T	A	T	G	G	46bp	C	A	T	A	A	Type1
9311	G	A	G	C	C	C	A	C	A	A	G	C	G	T	G	G	46bp	C	G	C	G	G	Type2
Chenghui9348	G	A	C	C	A	A	G	A	G	G	A	C	G	C	A	A	-	T	G	C	G	G	Type3

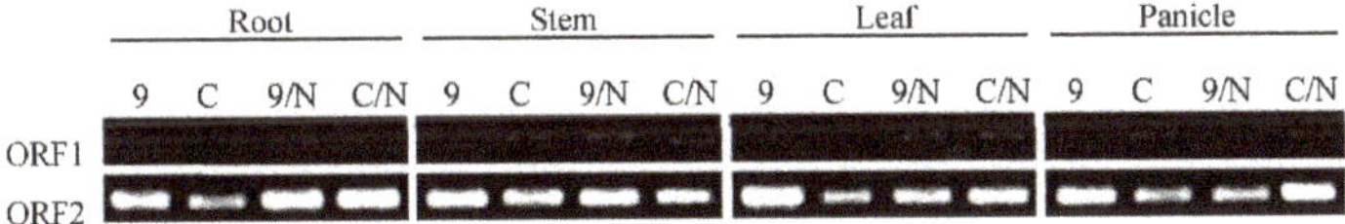

ORF2

Position Cultivars	17	98	137	166	253	285	348	390	489	501	527	543	558	587-598	602	603	614	653	706	720	749	758	767	815	859	893	898	935	1007	1231	1286	1338	
Nipponbare	G	A	C	T	A	A	C	T	G	T	T	C	T	-	C	C	C	G	A	T	C	G	A	G	G	T	G	G	G	C	C	T	Type1
9311	G	A	G	G	C	G	C	T	A	T	C	C	T	-	T	G	T	T	A	T	C	G	A	A	G	T	C	G	A	A	C	C	Type2
Chenghui9348	A	G	G	G	C	G	T	C	A	C	C	T	C	12bp	T	G	T	T	T	C	T	A	C	A	A	C	C	T	A	A	T	C	Type3
	A	G	C	G	C	G	T	C	G	C	C	C	C	12bp	T	G	T	G	A	C	C	G	C	T	G	C	G	T	G	C	C	T	Type4

Figure 5. Different nucleotide sequences of ORF1 and ORF2 in the 38 cultivars. (**a**) Green strips show the exons of ORF1, and the sequence features as indicated. (**b**) Blue strips show the exons of ORF2, and the sequence features as indicated. SNPs and insertion/deletion positions with ORF1 and ORF2 are connected by lines.

3.7. mRNA Expression of Candidate Genes of F12-C

The expression patterns of ORF1 and ORF2 in parents (9311, Chenghui9348) and F_1 hybrids (9311/Nipponbare, Chenghui9348/Nipponbare) were analyzed by RT-PCR. The results suggested that ORF1 was expressed constitutively in many organs such as the mature root, mature stem, mature leaf, and panicles. Unfortunately, we did not observe any significant difference between the F_1 of the Chenghui9348/Nipponbare and 9311 × Nipponbare hybrids in various tissues (Figure 6). ORF2, however, did not show any expression in all examined tissues.

Figure 6. Expression level of ORF1 in various tissues. 9, 9311. C, Chenghui9348. 9/N, the F_1 of 9311/Nipponbare. C/N, the F_1 of Chenghui9348/Nipponbare.

4. Discussion

We have identified a new wide-compatibility locus, *F12-C*, from Chenghui9348 in this study. Correspondingly, we designate the genotype of Chenghui9348 at this locus as *F12-C*, that of 9311 as *F12-9*, and that of Nipponbare as *F12-ni*. Consequently, individuals with *F12-C/F12-9* and *F12-C/F12-ni* genotypes produced fertile pollen and panicles, while individuals with the *F12-9/F12-ni* genotype produced semi-sterile pollen and panicles due to allele interaction. There were no wide-compatibility genes reported in this region before; we considered *F12-C* as a new wide-compatibility locus that can improve the fertility of spikelets by improving pollen fertility in rice inter-subspecific hybrids.

A notable feature of the *F12-C* locus observed in this study is the obvious effect of male fertility on spikelet fertility. Just like the *F12-C* locus, there was only one *f5-Du* locus that played a similar role in a previous study [20,21]. *f5-Du*, which is a neutral allele, could increase pollen fertility by more than 50% and spikelet fertility by over 20%, and up to 79% and 35%, respectively [20]. *F12-C*, however, has an advantage compared with the *f5-Du* locus because it can increase spikelet fertility to approximately 70%. These results indicated that the *F12-C* locus has more potential in rice breeding programs. Apart from the *F12-C*, *f5-Du*, and *S5-n* loci, there are also some other wide-compatibility loci that have been found in different varieties, such as S1-g, S7-n, S8-n, S9-n, S15-n, S16-n, S29-n, S30-n, and S32-n, all of which were just found by using the position from a testcross [15,22–25].

Two predicted candidate genes were identified within the *F12-C* locus by comparing their sequences between parental lines. The deletion/insertion of these two genes may lead to loss of function in protein–protein interactions, resulting in normal fertility in hybrids with *indica* and *japonica*. Nowadays, several WCGs have been cloned and all of them have a deletion or insertion compared with the *indica* or *japonica* allele. The wide-compatibility allele S5-n has a 136-bp deletion at the N-terminus of the predicted protein, resulting in the loss of the signal peptide and therefore the mislocalization of the protein [11,12]. A second major locus, Sa, has been cloned in chromosome 1 and Dular carries a neutral allele at this locus. Sequence analysis of the Sa locus showed that Dular carries a 6-bp insertion compared with SaM in the *indica* and *japonica* allele, which is considered responsible for its fertility-neutral function [26,27]. Another neutral S4-n allele, namely *f5-Du*, has a 30-bp insertion near the 3'UTR of ANK-3 in Dular (AK105314) [21,28,29]. All of these studies suggested that the deletion/insertion of WCGs is responsible for their fertility-neutral function.

We found a high degree of polymorphism in the ORF1 and ORF2 sequences. The ORF1 allele in Chenghui9348 was conserved compared with ORF2; however, we could not confirm that ORF1 is *F12-C*, regardless of ORF2. It is possible that these two candidate genes that we deduced from the sequence characteristics of WCGs might not be *F12-C*. However, the candidate gene should be in the region between the D1101 and D1164 markers according to the results of the map-based cloning. To further clone the *F12-C* locus, larger mapping populations and smaller location regions, as well as complementation tests, are necessary.

The finding of *F12-C* might have significant implications in rice breeding programs. The wide-compatibility gene S5-n has been widely applied in many rice breeding programs to overcome the sterility of inter-subspecific hybrids. However, it has been frequently found that improvements in embryo sac fertility by the S5-n gene alone are not sufficient for producing *indica/japonica* hybrids with normal fertility [24,30–32]. So, it is necessary to explore other wide-compatibility genes, such as *f5-Du* and *F12-C*, to overcome hybrid sterility by increasing pollen fertility [20]. Nowadays, this can be easily achieved with marker-assisted selection using the markers identified in the present study.

5. Conclusions

In this study, we identified a wide-compatibility locus, *F12*, in inter-subspecific hybrids of rice and developed a pair of InDel markers for map-based cloning of *F12*. Moreover, two candidate genes, ORF1 and ORF2, were identified at the *F12* locus.

Supplementary Materials: The following supporting information can be downloaded at: https://www.mdpi.com/article/10.3390/agronomy12112851/s1, Table S1: Pollen and spikelet fertilities of F$_1$ hybrids between different varieties; Table S2: Primers used for mapping of the *F12-C* locus and re-sequencing are listed; Table S3: Annotation data of the putative genes in the *F12-C* region in the Nipponbare genome database from Gramene; Table S4: Nucleotide variation within the ORF1 and ORF2 sequences in a panel of 16 *indica* cultivars and 22 *japonica* cultivars.

Author Contributions: W.Z. (Weibo Zhao) analyzed the data; W.Z. (Wei Zhou) performed the experiments and hybridization; W.H. designed the research; J.F., H.G., Z.D., Y.Z., W.X. and Z.H. helped revise the manuscript; W.Z. (Weibo Zhao) and W.H. wrote the manuscript. All authors have read and agreed to the published version of the manuscript.

Funding: This research was funded by the National Key R&D Program of China (grant No. 2017YFD0100 400), the Creative Research Groups of the Natural Science Foundation of Hubei Province (2020CFA009), the National Natural Science Foundation of China (grant No. 31771746), and the National Rice Industry Technology System (grant No. CARS-01-07).

Data Availability Statement: The sequences of genes and proteins mentioned in our study are available for download from the public database mentioned above.

Conflicts of Interest: The authors declare no conflict of interest.

References

1. Xie, Y.; Shen, R.; Chen, L.; Liu, Y.G. Molecular mechanisms of hybrid sterility in rice. *Sci. China Life Sci.* **2019**, *62*, 737–743. [CrossRef] [PubMed]
2. Britwum, K.; Demont, M. Tailoring rice varieties to consumer preferences induced by cultural and colonial heritage: Lessons from New Rice for Africa (NERICA) in The Gambia. *Outlook Agric.* **2021**, *50*, 305–314. [CrossRef] [PubMed]
3. Kato, A. On the affinity of rice varieties as shown by the fertility of rice plants. *Cent. Agric. Inst. Kyushu Imp. Univ.* **1928**, *2*, 241–276.
4. Ikehashi, H.; Araki, H. Varietal Screening of Compatibility Types Revealed in F1 Fertility of Distant Crosses in Rice. *Jpn. J. Breed.* **1984**, *34*, 304–313. [CrossRef]
5. Kitamura, E. Genetic studies on sterility observed n hybrids between distantly related varieties of rice. *Jpn. J. Breed.* **1962**, *12*, 166–168. [CrossRef]
6. Oka, H.I. Analysis of Genes Controlling F1 Sterility in Rice by the Use of Isogenic Lines. *Genetics* **1974**, *77*, 521–534. [CrossRef] [PubMed]
7. Michelmore, R.W.; Paran, I.; Kesseli, R.V. Identification of markers linked to disease-resistance genes by bulked segregant analysis: A rapid method to detect markers in specific genomic regions by using segregating populations. *Proc. Natl. Acad. Sci. USA* **1991**, *88*, 9828–9832. [CrossRef]
8. Yanagihara, S.; McCouch, S.R.; Ishikawa, K.; Ogi, Y.; Maruyama, K.; Ikehashi, H. Molecular analysis of the inheritance of the S-5 locus, conferring wide compatibility in Indica/Japonica hybrids of rice (*O. sativa* L.). *TAG Theor. Appl. Genet. Theor. Angew. Genet.* **1995**, *90*, 182–188. [CrossRef]
9. Ji, Q.; Lu, J.; Chao, Q.; Gu, M.; Xu, M. Delimiting a rice wide-compatibility gene S5n to a 50 kb region. *TAG Theor. Appl. Genet. Theor. Angew. Genet.* **2005**, *111*, 1495–1503. [CrossRef]
10. Qiu, S.Q.; Liu, K.; Jiang, J.X.; Song, X.; Xu, C.G.; Li, X.H.; Zhang, Q. Delimitation of the rice wide compatibility gene S5 (n) to a 40-kb DNA fragment. *TAG Theor. Appl. Genet. Theor. Angew. Genet.* **2005**, *111*, 1080–1086. [CrossRef]
11. Chen, J.; Ding, J.; Ouyang, Y.; Du, H.; Yang, J.; Cheng, K.; Zhao, J.; Qiu, S.; Zhang, X.; Yao, J.; et al. A triallelic system of S5 is a major regulator of the reproductive barrier and compatibility of indica-japonica hybrids in rice. *Proc. Natl. Acad. Sci. USA* **2008**, *105*, 11436–11441. [CrossRef] [PubMed]
12. Yang, J.; Zhao, X.; Cheng, K.; Du, H.; Ouyang, Y.; Chen, J.; Qiu, S.; Huang, J.; Jiang, Y.; Jiang, L.; et al. A killer-protector system regulates both hybrid sterility and segregation distortion in rice. *Science* **2012**, *337*, 1336–1340. [CrossRef] [PubMed]
13. Liu, K.D.; Zhou, Z.Q.; Xu, C.G.; Zhang, Q.F.; Maroof, M.A.S. An analysis of hybrid sterility in rice using a diallel cross of 21 parents involving indica, japonica and wide compatibility varieties. *Euphytica* **1996**, *90*, 275–280. [CrossRef]
14. Wan, J.M.; Yanagihara, S.; Kato, H.; Ikehashi, H. Multiple Alleles at a New Locus Causing Hybrid Sterility between a Korean Indica Variety and a Javanica Variety in Rice (*Oryza-Sativa* L). *Jpn. J. Breed.* **1993**, *43*, 507–516.
15. Wan, J.; Yamaguchi, Y.; Kato, H.; Ikehashi, H. Two new loci for hybrid sterility in cultivated rice (*Oryza sativa* L.). *TAG Theor. Appl. Genet. Theor. Angew. Genet.* **1996**, *92*, 183–190. [CrossRef] [PubMed]
16. Liu, K.D.; Wang, J.; Li, H.B.; Xu, C.G.; Liu, A.M.; Li, X.H.; Zhang, Q. A genome-wide analysis of wide compatibility in rice and the precise location of the S5 locus in the molecular map. *TAG Theor. Appl. Genet. Theor. Angew. Genet.* **1997**, *95*, 809–814. [CrossRef]
17. Ouyang, Y.; Liu, Y.G.; Zhang, Q. Hybrid sterility in plant: Stories from rice. *Curr. Opin. Plant Biol.* **2010**, *13*, 186–192. [CrossRef]
18. Zhou, S.; Wang, Y.; Li, W.; Zhao, Z.; Ren, Y.; Wang, Y.; Gu, S.; Lin, Q.; Wang, D.; Jiang, L.; et al. Pollen semi-sterility1 encodes a kinesin-1-like protein important for male meiosis, anther dehiscence, and fertility in rice. *Plant Cell* **2011**, *23*, 111–129. [CrossRef]
19. McCouch, S.R.; Teytelman, L.; Xu, Y.; Lobos, K.B.; Clare, K.; Walton, M.; Fu, B.; Maghirang, R.; Li, Z.; Xing, Y.; et al. Development and mapping of 2240 new SSR markers for rice (*Oryza sativa* L.). *DNA Res.* **2002**, *9*, 199–207. [CrossRef]
20. Wang, G.W.; He, Y.Q.; Xu, C.G.; Zhang, Q. Identification and confirmation of three neutral alleles conferring wide compatibility in inter-subspecific hybrids of rice (*Oryza sativa* L.) using near-isogenic lines. *TAG Theor. Appl. Genet. Theor. Angew. Genet.* **2005**, *111*, 702–710. [CrossRef]
21. Wang, G.W.; He, Y.Q.; Xu, C.G.; Zhang, Q. Fine mapping of f5-Du, a gene conferring wide-compatibility for pollen fertility in inter-subspecific hybrids of rice (*Oryza sativa* L.). *TAG Theor. Appl. Genet. Theor. Angew. Genet.* **2006**, *112*, 382–387. [CrossRef] [PubMed]
22. Zhu, S.S.; Jiang, L.; Wang, C.M.; Zhai, H.Q.; Li, D.T.; Wan, J.M. The origin of weedy rice Ludao in China deduced by genome wide analysis of its hybrid sterility genes. *Breed. Sci.* **2005**, *55*, 409–414. [CrossRef]
23. Hu, F.Y.; Xu, P.; Deng, X.N.; Zhou, J.W.; Li, J.; Tao, D.Y. Molecular mapping of a pollen killer gene S29(t) in Oryza Glaberrima and co-linear analysis with S22 in O-Glumaepatula. *Euphytica* **2006**, *151*, 273–278. [CrossRef]
24. Li, D.; Chen, L.; Jiang, L.; Zhu, S.; Zhao, Z.; Liu, S.; Su, N.; Zhai, H.; Ikehashi, H.; Wan, J. Fine mapping of S32(t), a new gene causing hybrid embryo sac sterility in a Chinese landrace rice (*Oryza sativa* L.). *TAG Theor. Appl. Genet. Theor. Angew. Genet.* **2007**, *114*, 515–524. [CrossRef]
25. Deng, X.; Zhou, J.; Xu, P.; Li, J.; Hu, F.; Tao, D. The role of *S1-g* allele from *Oryza glaberrima* in improving interspecific hybrid sterility between *O. sativa* and *O. glaberrima*. *Breed. Sci.* **2010**, *60*, 342–346. [CrossRef]
26. Long, Y.; Zhao, L.; Niu, B.; Su, J.; Wu, H.; Chen, Y.; Zhang, Q.; Guo, J.; Zhuang, C.; Mei, M.; et al. Hybrid male sterility in rice controlled by interaction between divergent alleles of two adjacent genes. *Proc. Natl. Acad. Sci. USA* **2008**, *105*, 18871–18876. [CrossRef]

27. Wang, Y.; Zhong, Z.Z.; Zhao, Z.G.; Jiang, L.; Bian, X.F.; Zhang, W.W.; Liu, L.L.; Ikehashi, H.; Wan, J.M. Fine mapping of a gene causing hybrid pollen sterility between Yunnan weedy rice and cultivated rice (*Oryza sativa* L.) and phylogenetic analysis of Yunnan weedy rice. *Planta* **2010**, *231*, 559–570. [CrossRef]

28. Kubo, T.; Yoshimura, A.; Kurata, N. Hybrid male sterility in rice is due to epistatic interactions with a pollen killer locus. *Genetics* **2011**, *189*, 1083–1092. [CrossRef]

29. Zhao, Z.G.; Zhu, S.S.; Zhang, Y.H.; Bian, X.F.; Wang, Y.; Jiang, L.; Liu, X.; Chen, L.M.; Liu, S.J.; Zhang, W.W.; et al. Molecular analysis of an additional case of hybrid sterility in rice (*Oryza sativa* L.). *Planta* **2011**, *233*, 485–494. [CrossRef]

30. Li, W.; Zeng, R.; Zhang, Z.; Ding, X.; Zhang, G. Identification and fine mapping of S-d, a new locus conferring the partial pollen sterility of intersubspecific F1 hybrids in rice (*Oryza sativa* L.). *TAG Theor. Appl. Genet. Theor. Angew. Genet.* **2008**, *116*, 915–922. [CrossRef]

31. Chen, M.J.; Zhao, Z.G.; Jiang, L.; Wan, J.M. A new gene controlling hybrid sterility in rice (*Oryza sativa* L.). *Euphytica* **2012**, *184*, 15–22. [CrossRef]

32. Zhao, J.Y.; Li, J.; Xu, P.; Zhou, J.W.; Hu, F.Y.; Deng, X.N.; Deng, W.; Tao, D.Y. A new gene controlling hybrid sterility between Oryza sativa and Oryza longistaminata. *Euphytica* **2012**, *187*, 339–344. [CrossRef]

Article

OsChlC1, a Novel Gene Encoding Magnesium-Chelating Enzyme, Affects the Content of Chlorophyll in Rice

Wei Lu [1], Yantong Teng [2], Fushou He [3], Xue Wang [3], Yonghua Qin [1], Gang Cheng [1], Xin Xu [1], Chuntai Wang [1] and Yanping Tan [1,*]

[1] Hubei Provincial Key Laboratory for Protection and Application of Special Plants in Wuling Area of China, College of Life Science, South-Central Minzu University, Wuhan 430074, China
[2] Biotechnology Research Institute, Chinese Academy of Agricultural Sciences, Beijing 100081, China
[3] College of Life Science, South-Central Minzu University, Wuhan 430074, China
* Correspondence: yanptan@mail.scuec.edu.cn

Abstract: Leaf-color mutants in rice (*Oryza sativa* L.) are excellent models for studying chlorophyll biosynthesis and chloroplast development. In this study, a yellow-green-leaf mutant generated by ^{60}Co irradiation, *ygl9311*, was isolated: it displayed a yellow-green leaf phenotype during the complete growth cycle. Compared with the wild type, the photosynthetic pigment contents of leaves in *ygl9311* were significantly reduced, and chloroplast development was retarded. Genetic analysis indicated that the *ygl9311* phenotype was controlled by a single recessive nuclear gene. Map-based cloning and transcriptome sequencing analysis suggested that the candidate gene was *OsChlC1* (*BGIOSGA012976*), which encodes a Mg-chelatase I subunit. The results of CRISPR/Cas9 system and RNAi knockout tests show that mutation of *OsChlC1* could reproduce the phenotype of yellow-green leaves of the mutant *ygl9311*. In conclusion, the novel rice leaf-color gene *OsChlC1* affects the content of chlorophyll in rice, showing a relatively conserved function in *indica* and *japonica* rice cultivars.

Keywords: rice; leaf-color mutants; chlorophyll; *ygl9311*; *OsChlC1*; Mg-chelatase; OsChlI subunit

Citation: Lu, W.; Teng, Y.; He, F.; Wang, X.; Qin, Y.; Cheng, G.; Xu, X.; Wang, C.; Tan, Y. *OsChlC1*, a Novel Gene Encoding Magnesium-Chelating Enzyme, Affects the Content of Chlorophyll in Rice. *Agronomy* **2023**, *13*, 129. https://doi.org/10.3390/agronomy13010129

Academic Editor: Guodong Lu

Received: 24 November 2022
Revised: 27 December 2022
Accepted: 28 December 2022
Published: 30 December 2022

1. Introduction

Leaf-color mutants are often generated due to disorders in chlorophyll (Chl) metabolism during plant development [1–3]. The molecular mechanism of leaf-color mutant generation remains unclear because Chl metabolism involves complex biochemical reactions [4]. Rice (*Oryza sativa* L.) is an important food crop and also one of the most widely studied model plants. For this reason, many scholars have studied rice leaf-color mutants.

So far, hundreds of rice leaf-color mutants have been identified, and more than 30 leaf-color mutant genes have been cloned [5,6]. These genes are involved mostly in the metabolic pathway of photosynthetic pigments and can be assigned into three categories. The first category refers to genes that are directly involved in the synthesis of photosynthetic pigments and whose mutation hinders this synthesis. For instance, *OsCHLH* is a gene encoding the H subunit of magnesium ion (Mg^{2+}) chelatase during Chl synthesis, and its complete knockout kills mutants [7]. *OsCAO* encodes Chl oxygenase, and its mutation leads to the delayed development of rice plants and grayish-green leaves [8]. *YGL1* encodes Chl synthase, and its mutation makes rice leaves yellow-green in the seedling stage but normal green in the grain-filling stage [9]. *OsDVR*, a gene encoding α-8-vinyl reductase, catalyzes the reduction of divinyl Chl to vinyl Chl [10]. *YGL22* encodes a chloroplast protein, and its knockout results in the yellow-green leaf phenotype in the seedling stage [11]. The second category comprises genes involved mainly in the degradation of Chl and whose mutation makes rice leaves keep the evergreen phenotype. For example, *SGR* mutations hinder the degradation of Chl and pigment-related proteins in rice [12]. *NYC1* encodes Chl *b* reductase, and its mutation represses the degradation of chloroplast grana allowing rice leaves to remain green at harvest time [13,14]. *NYC3* mediates the degradation of the light-harvesting chromoprotein composites [15]. *NYC4* also participates in the degradation of Chl-protein composites. The third category represents genes that are involved mainly in the synthesis of photosynthetic pigments other than Chl [16]. For instance, OsPDS and β-OsLCY are key enzymes for the synthesis of carotenoid precursors [17].

Chl synthesis comprises two parts: biosynthesis of protoporphyrin IX from L-glutamyl-tRNA and then synthesis of Chl *a* and Chl *b* from protoporphyrin IX [18–21]. Mg-chelatase is one of the key rate-limiting enzymes for the synthesis of Chl from protoporphyrin IX, chelating Mg^{2+} to protoporphyrin IX to generate magnesium protoporphyrin IX [18]. Mg-chelatase is composed of three different subunits, namely CHLI (38–46 kDa), CHLD (60–87 kDa), and CHLH (140–155 kDa) [22–24]. Mg-chelatase has been proved in in vivo and in vitro recombination experiments to play its role only when all three subunits are present [22,25]. It is speculated that Mg-chelatase works through the following mechanism: Firstly, the CHLI subunit possesses ATPase activity [26]. Then, the CHLD subunit contains a protein-binding domain able to form the CHLD/CHLI complex with the CHLI subunit [26]. Finally, with the energy from ATP hydrolysis, the CHLH subunit chelates Mg^{2+} to protoporphyrin IX [27]. However, a detailed molecular mechanism of Mg-chelatase remains to be clarified. More Mg-chelatase-related genes need to be cloned to accelerate this process.

In this study, a yellow-green leaf mutant (*ygl9311*) was identified, with leaves showing the yellow-green phenotype during the complete growth cycle. Leaves of *ygl9311* had a significantly decreased content of photosynthetic pigments in contrast with the wild type. Transmission electron microscopy (TEM) observations showed small and distorted chloroplasts in the leaves of *ygl9311*, and the grana and lamellae in chloroplasts became less abundant, with blurred granum lamellae. Moreover, the candidate gene (named *OsChlC1* in this study) of *ygl9311* was located in a 403.1 kb physical interval on chromosome 3 as determined by map-based cloning in this study. The analysis showed that *OsChlC1* encoded the subunit I of Mg-chelatase, playing a key role in the Chl synthesis in rice.

2. Materials and Methods

2.1. Plant Materials and Inherited Pattern Analysis

Previously, Teng et al. obtained a yellow-green leaf mutant (*ygl9311*) by ^{60}Co-γ-ray mutagenesis from normal green leaf *indica* cultivar 9311 [28]. Normal *indica* cultivar Gangzao was crossed with mutant *ygl9311* to obtain the F_1 generation, and the F_2 population for gene mapping was obtained by selfing the F_1 plants.

The reciprocal cross experiment between mutant *ygl9311* and the *indica* cultivar Gangzao was carried out to study the genetic pattern of the yellow leaf mutant. The reciprocal cross between the cultivar 9311 and the *indica* cultivar Gangzao was taken as control. The phenotypes of the F_1 and F_2 generations were investigated, and the trait segregation ratio in F_2 generation was calculated and analyzed by chi square test.

All rice materials were planted under natural field conditions in either Wuhan, Hubei Province, China, or Lingshui, Hainan Province, China.

2.2. Measurement of Major Agronomic Traits

The mutant *ygl9311* and *indica* cultivar 9311 were planted to measure the major agronomic traits. After maturation, plant height, number of effective tillers, number of productive panicles per plant, and number of spikelets per panicle were recorded. Data of the mutant *ygl9311* and cultivar 9311 were analyzed in Microsoft Excel, and significance of difference between them was assessed using the *t*-test.

2.3. Quantitative Analysis of Chlorophyll Content

For pigment extraction, fresh leaves (second from the top) from three biological replicates were taken in the three-leaf stage, middle of tillering stage, and flowering stage. An amount of 0.15 g of fresh leaf per sample was homogenized and centrifuged in ice-cold 80% *v/v* acetone. The concentrations of chlorophyll *a* and *b* and carotenoids were calculated by measuring the absorbance of the supernatant at wavelengths 663 nm, 645 nm, and 470 nm, respectively, using the equation of Lichtenthaler [29]. All procedures were carried out under dim green light to avoid the degradation of photosynthetic pigments.

2.4. Observations of Chloroplast Structure

Leaf samples of the mutant *ygl9311* and *indica* cultivar 9311 were harvested from the second leaf from the top of seedlings in the 1-core 2-leaf stage. The samples were fixed with 0.1 mol/L phosphate buffer (pH 7.2) containing 3% *w/v* glutaraldehyde for 4 h. They were post-fixed with 1% *v/v* osmium acid (pH 7.2) for 4 h. The samples were dehydrated with 30%, 50%, 70%, 80%, 90%, and 100% acetone and then embedded in epoxy resin SPURR. The samples were cut into approximately 1 μm thick sections with a freezing microtome. After staining, the ultrathin sections were observed under a transmission electron microscope (H-600IV, Hitachi, Tokyo, Japan).

2.5. Measurement of Total SOD Activity

Total superoxide dismutase (SOD) activity of the mutant *ygl9311* and *indica* cultivar 9311 was measured with an SOD enzyme activity assay kit from Nanjing Jiancheng Company. The kit used the xanthine oxidase method to determine the activity of SOD. Five biological replicates, each containing 0.1 g of fresh leaves from a mature plant, were analyzed.

2.6. Initial Mapping of the Mutant ygl9311 Locus with Genome Re-Sequence

Mixed-genome DNA pools of mutant *ygl9311* and *indica* cultivar 9311 were built, consisting of 100 individual genome DNA. Re-sequence of the two pools was carried out on an Illumina Hiseq2000. The re-sequence data were aligned with the *japonica* reference genomic sequence (EnsemblPlants, http://plants.ensembl.org/Oryza_sativa/Info/Index, accessed on 21 May 2018) to remove low-quality data. Softs of GATK and annovar were used to align and annotate SNPs and small indels. Then, initial mapping was conducted by screening homozygous mutation sites where the mutant *ygl9311* was not consistent with *indica* cultivar 9311.

2.7. Fine Mapping of the Mutant ygl9311 Locus and Prediction of Candidate Genes

A total of 1325 recessive F_2 plants derived from more than 5300 plants of (the mutant *ygl9311*/*indica* cultivar Gangzao) F_2 population were selected for fine mapping. According to the result of the initial mapping, seven simple sequence repeat (SSR) markers (Table 1) located on chromosome 3 were used for making the linkage map for *ygl9311* locus. Marker information was provided from a web server, Gramene (http://www.gramene.org/bd/markers/, accessed on 5 June 2018). Transcriptome data between mutant *ygl9311* and *indica* cultivar 9311 were used to predict candidate genes. Those genes located within the mapping interval and expressed differentially were considered as candidate genes. The function of candidate genes was predicted with EnsemblPlants.

Table 1. Primer sequences of SSR markers used in the mapping of *ygl9311* locus.

SSR Markers	Forward Primer (5′-3′)	Reverse Primer (5′-3′)
RM15177	TCCTGTGTTGGACGGAGTATGC	GCCTCAGAGGTTAGAAGACAGACAGC
RM15189	GGTATCTCCCAGACACACATAGTGG	GATTGTCTCCATATCTCAGCATCC
RM15217	AAGAACCCACCTGCGGTTAGC	CTACAGCTTTCTTGATTCGCTTGG
RM15245	AGGATTTACACGCGCTTTGAGC	CATCAACGGCAGTAGAAGGTTTCC
RM15303	GAATCGGGTCTACGGTTTAGG	AAAGGAAGAGAAGAGGCAACG
RM15313	GATAAGGACATGGTGTGGTCACG	GGCCAACTAAGCACAACAAATACC
RM15355	GTAGGAAATTCTTCGCCAGATGC	CCGAGACTTGGAACAATCTTAGGC

2.8. Validation of the Function of Candidate Gene

To validate the function of candidate gene (*OsChlC1*), the knockout construct of *OsChlC1* was built through the gene-editing vector CRISPR/Cas9 [30]. Two specific editing fragments were amplified with the primers F1 5′-GGCAGCACCGTGTCGAGCGCGTA-3′ and 5′-AAACTACGCGCTCGACACGGTGC-3′ and F2 5′-GGCAGCCGCCGTGTACCCTTCTA-3′ and 5′-AAACTAGAAGGGTACACGGCGGC-3′. The two specific editing sites were then inserted into the entry vector SK-gRNA. The gRNA-F1 and gRNA-F2 fragments were cloned into the binary vector pCAMBIA1300-Cas9. The resulting construct was designated as pCAMBIA1300-*Cas9-OsChlC1* and was transformed into japonica cultivar Nipponbare by *Agrobacterium*-mediated transformation. Primers 5′-GAATCCCTCAGCATTGTTC-3′ and 5′-TTGCGTCGTGCAGTCTGT-3′ were designed to detect the editing sites in the positive transgenic plants.

The RNAi construct of *OsChlC1* was also built in the vector DS1301. The specific interference fragment was amplified with the primers 5′-TAAACTAGTGGTACCGAACGCCTT-GACACCATCGG-3′ and 5′-TAAGAGCTCGGATCCAATCCCTTCGAGACTTGGGTG-3′. The forward primer contained a *Kpn*I and an *Spe*I site, and the reverse primer contained an *Sac*I and a *Bam*HI site. The resulting construct was designated as pDS1301-*OsChlC1*-RNAi and was transformed into *indica* cultivar 9311 by *Agrobacterium*-mediated transformation.

3. Results

3.1. Inheritance Pattern of the ygl9311 Mutant

The leaf-color mutant *ygl9311* in rice was induced by ^{60}Co-γ-ray mutagenesis from *indica* rice cultivar 9311; it had yellow-green leaves during the whole growth cycle (Figure 1). To analyze the genetic pattern of the mutant *ygl9311*, direct and reciprocal crosses with the mutant *ygl9311* and *indica* rice cultivar Gangzao as parents were conducted in this study. It was found that the leaves of both direct and reciprocal cross F_1 generations were all green (Figure 1B), but both direct and reciprocal cross F_2 generations showed trait segregation of yellow-green and normal green leaves. According to statistics, 926 seedlings with green leaves and 301 seedlings with yellow-green leaves were found in the direct cross F_2 generation, and 1087 green seedlings and 349 yellow-green seedlings were identified in the reciprocal cross F_2 generation. The segregation ratio of green leaf seedlings to yellow-green leaf seedlings in direct cross (926 green: 301 yellow-green, $x^2 = 0.144 < x^2_{0.05,1} = 3.84$) and reciprocal cross (1087 green: 349 yellow-green, $x^2 = 0.379 < x^2_{0.05,1} = 3.84$) F_2 generations is consistent with the Mendelian single-gene segregation rule of 3:1. These results indicate that the mutant *ygl9311* is a recessive inheritance controlled by a single gene.

Figure 1. Phenotypic characterization of the mutant *ygl9311*. (**A**) Plants at seedling stage. (**B**) Plants at mature stage grown in Hainan.

3.2. Characteristics of the Mutant ygl9311

Several groups of major agronomic traits of the mutant *ygl9311* and wild *indica* cultivar 9311 were assessed in this study to explore the influence of the mutation site on the agronomic traits of rice. When planted in Wuhan, Hubei Province, China, the plant height, number of productive panicles per plant, and number of spikelets per panicle of the mutant *ygl9311* were 31.1%, 53.2%, and 20.5%, respectively, lower than those of the wild-type *indica* cultivar 9311, but the number of effective tillers per plant showed no significant difference between the mutant *ygl9311* and the *indica* cultivar 9311 (Figure 2). However, when planted in Lingshui, Hainan Province, China, the mutant *ygl9311* grew well and displayed agronomic traits, such as the plant height, comparable to the *indica* cultivar 9311. The above results signify that the mutation site affected the agronomic traits of rice plants planted in Wuhan, but they exerted less effect on agronomic traits when rice plants were planted in Hainan.

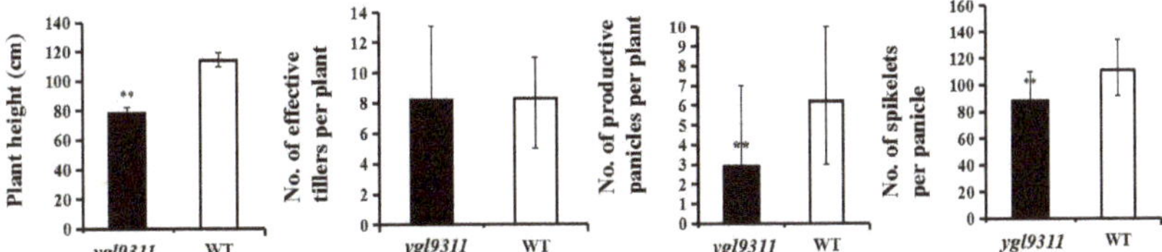

Figure 2. Comparison of major agronomic traits between mutant *ygl9311* and the wild-type parent 9311. Bars represent standard deviations of ten independent measurements. Significant differences were determined by Duncan's Multiple Range test.

The total superoxide dismutase (SOD) activity in the mutant *ygl9311* was detected to assess the effects of the mutation site on the stress resistance and antioxidant capacity of rice. The results indicate that the total SOD activity of mutant plants was 1474 U/g, slightly higher than that of the wild type (Figure 3), implying that the mutation site did not weaken the stress resistance and antioxidant capacity of rice plants.

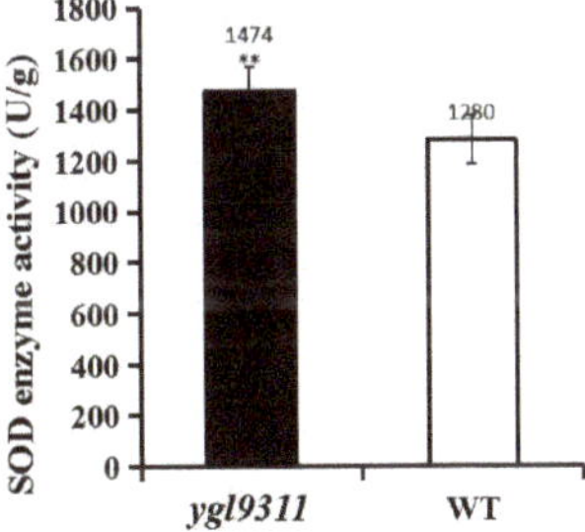

Figure 3. SOD enzyme activity in mutant *ygl9311* and the wild-type parent 9311. Bars represent standard deviations of three independent measurements. Significant differences were determined by Duncan's Multiple Range test.

Additionally, the photosynthetic pigment content in the leaves of the mutant *ygl9311* and *indica* cultivar 9311 was measured in this study at the one-core three-leaf stage, middle tillering stage, and heading stage, to investigate the physiological causes of the mutant *ygl9311*. At the above three stages, the content of Chl *a* in the mutant was, respectively, 58%, 80%, and 83% lower than that in the *indica* cultivar 9311, that of Chl *b* in the mutant was, respectively, 86%, 96%, and 99% lower than that in the *indica* cultivar 9311, and the carotenoid content in the mutant was, respectively, 58%, 77%, and 79% lower than that in the *indica* cultivar 9311 (Figure 4). The content of photosynthetic pigments was significantly lower in the mutant than in the *indica* cultivar 9311 in both the seedling stage and at maturity, consistent with the mutant keeping yellow-green leaves during the whole growth cycle. The above results demonstrate that the change in the leaf color of the mutant is attributed mainly to decreased pigment content in mutant plants.

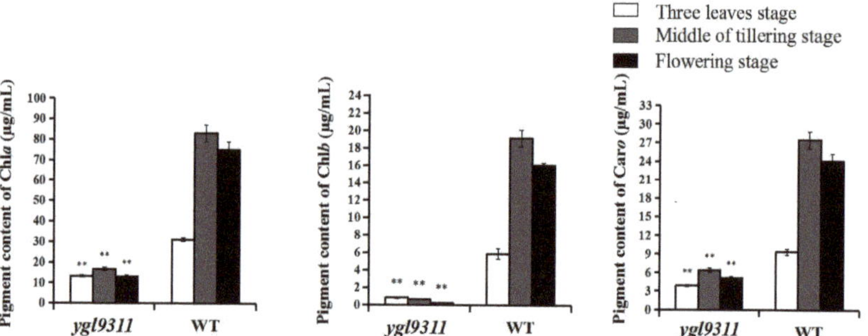

Figure 4. Photosynthetic pigment contents in leaves of the mutant *ygl9311* and the wild-type parent 9311. Bars represent standard deviations of three independent experiments. Significant differences were determined by Duncan's Multiple Range test.

To analyze further the effect of the mutation site on chloroplast development, the structure and morphology of chloroplasts in the mutant *ygl9311* and *indica* cultivar 9311 were examined by TEM (Figure 5). The results reveal that chloroplasts in the leaves of the mutant had a small size, distorted shape, and abnormal structure, in which the abundance of grana and lamellae declined, with blurred granum lamellae, fewer cristae, and obviously more osmiophilic granules. These results indicate that the mutant *ygl9311* has abnormal chloroplast development.

3.3. Map-Based Cloning of Locus of the Mutant ygl9311

In the present study, homozygous mutation sites between the mutant and the wild cultivar genome were screened out using the genome re-sequencing method to locate the mutation site of mutant *ygl9311*. According to re-sequencing results, 114,186,956 and 113,101,763 high-quality sequences with a reading length of over 100 bp were obtained in the mutant mixed pool and *indica* cultivar 9311 mixed pool, respectively. Next, the above sequences were processed with GATA software, and 4,820,964 single-nucleotide polymorphism (SNP) sites and 136,678 small InDel sites were identified. Importantly, homozygous mutation sites in the genome of the mutant were located on rice chromosomes. It was discovered that there was a SNP peak (at 20,546–21,231 Kb) on chromosome 3 (Figure 6). Therefore, it was speculated that the mutation site was located in the physical interval (20,546–21,231 Kb) on chromosome 3.

Thereafter, a F_2 population was constructed with the mutant *ygl9311* and *indica* cultivar Gangzao as parents to verify the above location results and obtain the fine-mapping of the mutant site. In brief, 1325 homozygous and recessive F_2 yellow-green leaf plants were screened using SSR primers within the initial mapping interval and in both flanking regions, on chromosome 3. The results show that the mutation site was located in the genetic interval of 7.77 cM on chromosome 3 based on seven SSR markers (Figure 7A). The *ygl9311* locus was finally narrowed to two SSR markers RM15303 and RM15313. We analyzed the sequences of these two SSR markers on the EnsemblPlants database and found that the mutation site was located at the physical interval (22,403,557–22,833,663 bp) on chromosome 3, between RM15303 (366.5 kb) and RM15313 (62.2 kb), in the genome of *indica* cultivar 9311. This physical interval has a length of 430.1 kb, covering 27 encoding genes (Figure 7B).

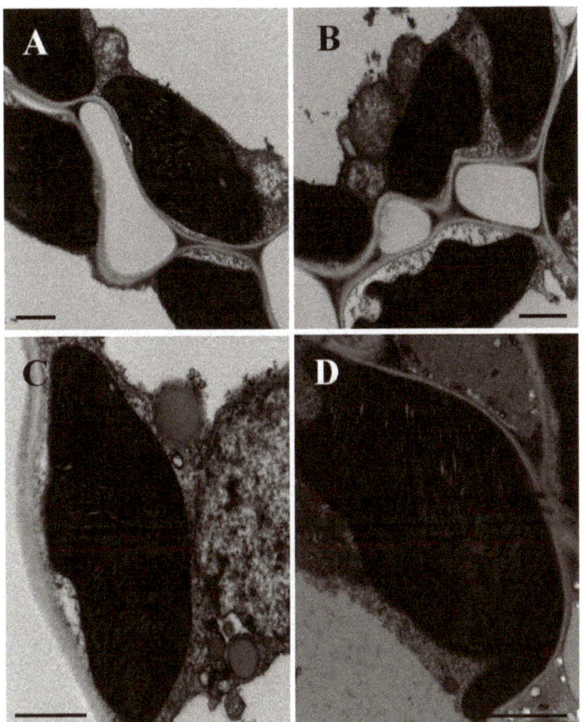

Figure 5. Electron microscope images of the mutant *ygl9311* and the wild-type plant at seedling stage. (**A,B**) Mesophyll cells of the wild type and *ygl9311*, respectively. (**C,D**) Chloroplasts of the wild type and *ygl9311*, respectively. Scale bar equals 1 μm.

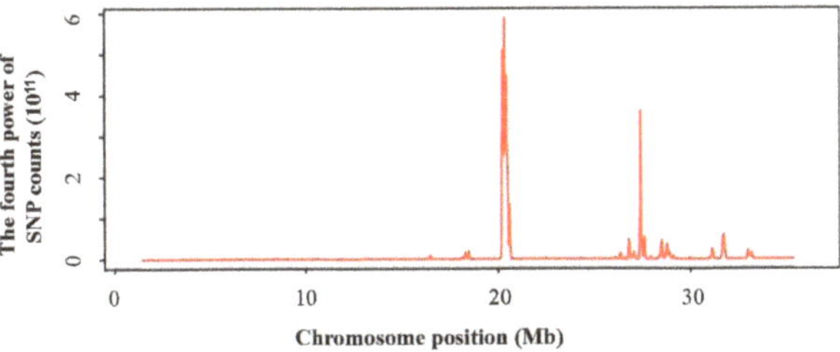

Figure 6. Distribution of SNP loci on chromosome 3. Peak at the position of 20,546–21,231 Kb on the chromosome 3 of *indica* cultivar 9311.

Transcriptome sequencing was performed between the mutant *ygl9311* and the *indica* cultivar 9311 in this study to determine the candidate gene of the mutation site. It was found that there were 217 differentially expressed genes (data not shown) between the mutant *ygl9311* and the *indica* cultivar 9311. Among these differentially expressed genes, two genes were in the physical interval of the yellow-green leaf mutation site (Table S1). Bioinformatics analysis manifested that among these differentially expressed genes, both *BGIOSGA012976* and *BGIOSGA010427* were encoding the Mg-chelatase I subunit and were implicated in chlorophyll content. However, only *BGIOSGA012976* showed sequence differences between *indica* cultivar 9311 and the mutant *ygl9311*. Therefore, it was considered in the present study that *BGIOSGA012976* was the candidate gene of the mutation *ygl9311*, and we named it *OsChlC1*.

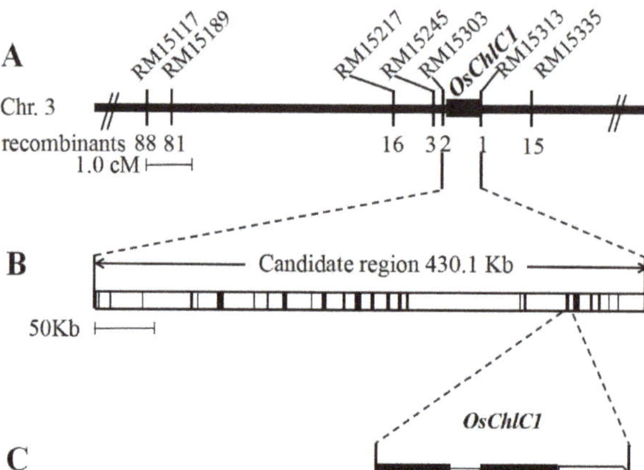

Figure 7. Molecular mapping of the *ygl9311* locus. (**A**) *OsChlC1* was narrowed between SSR markers RM15303 and RM15313 basing on analysis of 1325 recessive F$_2$ plants. (**B**) Total of 2 differentially expressed genes were found in the region between RM15303 and RM15313. (**C**) Candidate gene *OsChlC1* comprises two exons and one intron.

OsChlC1 is composed of two exons and one intron and has an open reading frame of 1152 bp, which can encode a polypeptide molecule consisting of 383 aa residues. In the present study, it was predicted that this gene has a chloroplast signal peptide (http://www.cbs.dtu.dk/services/TargetP/, accessed on 23 September 2020; Figure S1) at its N-terminus. According to sequence analysis, *OsChlC1* has a deletion mutation, and the first 640 bases of CDS of *OsChlC1* are absent in the genome of mutant *ygl9311*.

3.4. Validating the Function of OsChlC1

To validate the function of the candidate gene (*OsChlC1*), the knockout vector PC1300-*CAS9-OsChlC1* was constructed with the CRISPR-Cas9 technology, and genetic transformation was carried out on the *indica* cultivar 9311 with normal leaf color. Next, phenotypic identification was conducted on five transgenic rice plants that had *OsChlC1* successfully knocked out. It was uncovered that the leaf color of these rice plants with *OsChlC1* knocked out was yellow-green, consistent with that of the mutant *ygl9311* (Figure 8). Therefore, it was confirmed that the phenotype of yellow-green leaves of the mutant *ygl9311* was due to the mutation of the *OsChlC1* gene.

Figure 8. The phenotypes of the transgenic receptor parent *indica* 9311 (WT), mutant *ygl9311*, and knockout transgenic plant (KO) at seedling stage.

4. Discussion

In this study, the *indica* cultivar 9311 was subjected to mutation with ^{60}Co-γ-rays, and the mutant *ygl9311* with yellow-green leaves was then screened out. Next, the leaf-color gene was located by the map-based cloning strategy, cloned, and named *OsChlC1*. The results of knockout tests show that the mutant of *OsChlC1* could reproduce the phenotype of yellow-green leaves of the mutant *ygl9311*. Therefore, a new rice leaf-color gene *OsChlC1* was cloned in this study.

4.1. Chlorina Phenotype of Mutant ygl9311 Results from the Impaired Photosynthetic Pigment Synthesis

The contents of Chl *a*, Chl *b*, and carotene were significantly lower in the leaves of the mutant *ygl9311* than the *indica* cultivar 9311 in the seedling, tillering, and heading stages (Figure 4). Such a change in photosynthetic pigment content is consistent with the fact that the mutant has the traits of yellow-green leaves during the whole growth cycle. Under the catalysis of chlorophyllin a oxidase, chlorophyllin is converted into Chl *b*. The ratio Chl *a*/Chl *b* was significantly higher in the mutant *ygl9311* than the *indica* cultivar 9311 (Table S2), indicating an effect on the conversion of Chl *a* to Chl *b*. However, the ratio of total Chl content to carotene content exhibited no obvious difference between the two genotypes.

Photosynthesis-enabling protein complexes are embedded in the thylakoid membrane of chloroplasts [31–33]. The ultrastructural observation of chloroplasts revealed that the chloroplasts of the mutant *ygl9311* became smaller in size and distorted in shape, with an obviously decreased number of grana and lamellae (Figure 5). The above results indicate that the decrease in photosynthetic pigment content in the mutant *ygl9311* is the key factor leading to the phenotype of yellow-green leaves during the whole growth cycle. Moreover, the decrease in photosynthetic pigment content in the mutant *ygl9311* was associated with the defective development of chloroplasts.

4.2. The Mutation of OsChlC1 Results in Decreased Photosynthetic Pigment Content in ygl9311

Using a F_2 mapping population, the mutant gene was localized on chromosome 3, and its candidate gene *BGIOSGA012976* was cloned. The results of function prediction revealed that the gene encoded the Mg-chelatase I subunit; hence, the gene was named *OsChlC1*. The CRISPR-Cas9 technique was employed to create a mutant of the *indica* cultivar 9311 with *OsChlC1* knocked out. These transgenic *indica* rice plants showed the same yellow-green leaf phenotype as that of the yellow-leaf mutant (Figure 8). In addition, the knockout vector was transformed into the japonica rice cultivar Nipponbare, and these transgenic *japonica* rice plants also had the yellow-green leaf phenotype as that of the mutant *ygl9311* (Figure S2). The results of these two knockout experiments suggest a conserved function of *OsChlC1* in *indica* and *japonica* rice cultivars. Moreover, the knockout tests in both *indica* and *japonica* rice cultivars fully proved that *OsChlC1* mutation gives rise to the phenotype of yellow-green leaves.

4.3. OsChlC1 Is a Novel Mg-Chelatase Gene

As one of the key rate-limiting enzymes in chloroplast synthesis, Mg-chelatase binds Mg^{2+} to protoporphyrin IX to generate magnesium protoporphyrin IX [34,35]. Mg-chelatase is a complex protein composed of I, D, and H subunits [36]. Subunit I is a member of the AAA+ superfamily and has ATPase activity, whereas subunit D can form a binary complex with subunit I and interact with subunit H to insert Mg into protoporphyrin IX [37]. The Chl I subunit, an AAA+ ATPase, is responsible for the hydrolysis of ATP [38], with two very conserved domains, walker A and walker B, at its N-terminus, with walker A binding stably to ATP and walker B promoting the hydrolysis of ATP [39,40].

Many *CHLI* genes have been cloned in green plants. Two *CHLI* genes, namely *CHLI1* and *CHLI2*, exist in *Arabidopsis*. *AtCHLI1* mutation leads to the phenotype of gray-green leaves [41]. In rice, a point mutation on the third exon in the *OsCHLI* gene in the *Chlorina-9* mutant weakens the Mg-chelatase activity [42]. A follow-up study showed that in the *Chlorina-9* mutant, the CHLI subunit could not bind to the CHLD subunit to trigger the synthesis of Mg protoporphyrin IX. *OsCHLI* comprises three exons and two introns and encodes 415 Aa residues. The *Chlorina-9* mutant of rice has the phenotype of yellow-green leaves at the three-leaf stage, but its leaf color returns to normal green in the heading stage [42]. In the present study, *OsChlC1*, a new Mg-chelatase I subunit gene, was cloned in the mutant *ygl9311*. *OsChlC1* encoded 383 aa residues and was composed of exons and one intron. The mutant *ygl9311* exhibited the phenotype of yellow-green leaves during the whole growth cycle of rice. In conclusion, *OsChlC1* is a newly cloned Mg-chelatase I subunit gene in rice, and its mutation reduces the content of chlorophyll in rice, leading to the phenotype of yellow-green leaves.

Although a series of results were obtained in this study, there are still some studies to be further improved. We constructed the ectopic expression vector of the *OsChlC1* gene, and genetic transformation is already in progress. However, no positive plants have been obtained from the complementary assay. On the other hand, CRISPR/Cas9 system and RNAi knockout tests showed that the mutant of *OsChlC1* could reproduce the phenotype of yellow-green leaves of the mutant *ygl9311*, but we still need to determine the content of chlorophyll in the above mutant leaves. Gene function prediction indicates that *OsChlC1* encodes Mg-chelatase subunit I. In order to verify the function of *OsChlC1*, we need to detect the activity of magnesium ion chelatase in different mutants.

5. Conclusions

In this study, the candidate gene *OsChlC1* of the rice yellow-green leaf mutant *ygl9311* was cloned. According to the sequence analysis and function prediction, *OsChlC1* encoding 383 aa residues is a new Mg-chelatase I subunit gene. In addition, *OsChlC1* knockout in the *indica* rice cultivar 9311 and the *japonica* rice cultivar Nipponbare had the phenotype of yellow-green leaves of the mutant *ygl9311*. Therefore, we conclude that *OsChlC1* mutation in the genome of the *indica* rice cultivar 9311 leads to a decrease in chlorophyll content in rice leaves, affects chloroplast development, and results in the phenotype of yellow-green leaves of the mutant *ygl9311*. The *OsChlC1* exerts a relatively conserved function in *indica* and *japonica* rice cultivars.

Supplementary Materials: The following supporting information can be downloaded at https://www.mdpi.com/article/10.3390/agronomy13010129/s1, Figure S1: Chloroplast Signal Peptide Prediction of *OsChlC1* Gene, Figure S2: The Phenotypes of the wild-type parent 9311 (WT), mutant *ygl9311*, knockout transgenic plant (KO), and transgenic recipient parent Nipponbare (NIP) at seedling stage, Table S1: Differentially expressed genes within the physical location of the *ygl9311* mutation site, Table S2: Pigment content and ratio at various stages between *ygl9311* and wild type.

Author Contributions: Conceptualization, W.L., Y.T. (Yantong Teng) and F.H.; Data curation, W.L.; Investigation, G.C.; Methodology, W.L., Y.T. (Yantong Teng) and X.W.; Resources, C.W.; Software, Y.T. (Yantong Teng); Supervision, Y.T. (Yanping Tan); Validation, W.L., Y.Q. and X.X.; Writing—original draft, W.L.; Writing—review and editing, W.L. and Y.T. (Yanping Tan). All authors have read and agreed to the published version of the manuscript.

Funding: This research received no external funding.

Data Availability Statement: The datasets generated during the current study are available from the corresponding author on reasonable request.

Acknowledgments: The authors extend their appreciation for the support from the Hubei Provincial Key Laboratory for Protection and Application of Special Plants in Wuling Area of China, College of Life Science, South-Central Minzu University.

Conflicts of Interest: The authors declare no conflict of interest.

References

1. Khan, A.; Jalil, S.; Cao, H.; Tsago, Y.; Sunusi, M.; Chen, Z.; Shi, C.; Jin, X. The Purple Leaf (*pl6*) Mutation Regulates Leaf Color by Altering the Anthocyanin and Chlorophyll Contents in Rice. *Plants* **2020**, *9*, 1477. [CrossRef]
2. Wang, Y.; Wang, J.; Chen, L.; Meng, X.; Zhen, X.; Liang, Y.; Han, Y.; Li, H.; Zhang, B. Identification and function analysis of yellow-leaf mutant (*YX-yl*) of broomcorn millet. *BMC Plant Biol.* **2022**, *22*, 463. [CrossRef]
3. Chen, T.; Huang, L.; Wang, M.; Huang, Y.; Zeng, R.; Wang, X.; Wang, L.; Wan, S.; Zhang, L. Ethyl Methyl Sulfonate-Induced Mutagenesis and Its Effects on Peanut Agronomic, Yield and Quality Traits. *Agronomy* **2020**, *10*, 655. [CrossRef]
4. Yan, J.; Sun, P.; Liu, W.; Xie, D.; Wang, M.; Peng, Q.; Sun, Q.; Jiang, B. Metabolomic and Transcriptomic Analyses Reveal Association of Mature Fruit Pericarp Color Variation with Chlorophyll and Flavonoid Biosynthesis in Wax Gourd (*Benincasa hispida*). *Agronomy* **2022**, *12*, 2045. [CrossRef]
5. Deng, X.-J.; Zhang, H.-Q.; Wang, Y.; He, F.; Liu, J.-L.; Xiao, X.; Shu, Z.-F.; Li, W.; Wang, G.-H.; Wang, G.-L. Mapped clone and functional analysis of leaf-color gene *Ygl7* in a rice hybrid (*Oryza sativa* L. ssp. indica). *PLoS ONE* **2014**, *9*, e99564. [CrossRef]
6. Wan, C.; Li, C.; Ma, X.; Wang, Y.; Sun, C.; Huang, R.; Zhong, P.; Gao, Z.; Chen, D.; Xu, Z.; et al. *GRY79* encoding a putative metallo-β-lactamase-trihelix chimera is involved in chloroplast development at early seedling stage of rice. *Plant Cell Rep.* **2015**, *34*, 1353–1363. [CrossRef]
7. Jung, K.-H.; Hur, J.; Ryu, C.-H.; Choi, Y.; Chung, Y.-Y.; Miyao, A.; Hirochika, H.; An, G. Characterization of a rice chlorophyll-deficient mutant using the T-DNA gene-trap system. *Plant Cell Physiol.* **2003**, *44*, 463–472. [CrossRef]
8. Lee, S.; Kim, J.-H.; Yoo, E.-S.; Lee, C.-H.; Hirochika, H.; An, G. Differential regulation of chlorophyll a oxygenase genes in rice. *Plant Mol. Biol.* **2005**, *57*, 805–818. [CrossRef]
9. Wu, Z.; Zhang, X.; He, B.; Diao, L.; Sheng, S.; Wang, J.; Guo, X.; Su, N.; Wang, L.; Jiang, L.; et al. A chlorophyll-deficient rice mutant with impaired chlorophyllide esterification in chlorophyll biosynthesis. *Plant Physiol.* **2007**, *145*, 29–40. [CrossRef]
10. Wang, P.; Gao, J.; Wan, C.; Zhang, F.; Xu, Z.; Huang, X.; Sun, X.; Deng, X. Divinyl chlorophyll(ide) a can be converted to monovinyl chlorophyll(ide) a by a divinyl reductase in rice. *Plant Physiol.* **2010**, *153*, 994–1003. [CrossRef]
11. Zhu, Y.; Yan, P.; Dong, S.; Hu, Z.; Wang, Y.; Yang, J.; Xin, X.; Luo, X. Map-based cloning and characterization of *YGL22*, a new yellow-green leaf gene in rice (*Oryza sativa*). *Crop Sci.* **2020**, *61*, 529–538. [CrossRef]
12. Jiang, H.; Li, M.; Liang, N.; Yan, H.; Wei, Y.; Xu, X.; Liu, J.; Xu, Z.; Chen, F.; Wu, G. Molecular cloning and function analysis of the stay green gene in rice. *Plant J.* **2007**, *52*, 197–209. [CrossRef]
13. Kusaba, M.; Ito, H.; Morita, R.; Iida, S.; Sato, Y.; Fujimoto, M.; Kawasaki, S.; Tanaka, R.; Hirochika, H.; Nishimura, M.; et al. Rice NON-YELLOW COLORING1 is involved in light-harvesting complex II and grana degradation during leaf senescence. *Plant Cell* **2007**, *19*, 1362–1375. [CrossRef]
14. Sato, Y.; Morita, R.; Katsuma, S.; Nishimura, M.; Tanaka, A.; Kusaba, M. Two short-chain dehydrogenase/reductases, NON-YELLOW COLORING 1 and NYC1-LIKE, are required for chlorophyll b and light-harvesting complex II degradation during senescence in rice. *Plant J.* **2007**, *57*, 120–131. [CrossRef]
15. Morita, R.; Sato, Y.; Masuda, Y.; Nishimura, M.; Kusaba, M. Defect in *non-yellow coloring 3*, an α/β hydrolase-fold family protein, causes a stay-green phenotype during leaf senescence in rice. *Plant J.* **2009**, *59*, 940–952. [CrossRef]
16. Yamatani, H.; Sata, Y.; Masuda, Y.; Kato, Y.; Morita, R.; Fukunaga, K.; Nagamura, Y.; Nishimura, M.; Sakamoto, W.; Tanaka, A.; et al. *NYC4*, the rice ortholog of Arabidopsis *THF1*, is involved in the degradation of chlorophyll-protein complexes during leaf senescence. *Plant J.* **2013**, *74*, 652–662. [CrossRef]
17. Fang, J.; Chai, C.; Qian, Q.; Li, C.; Tang, J.; Sun, L.; Huang, Z.; Guo, X.; Sun, C.; Liu, M.; et al. Mutations of genes in synthesis of the carotenoid precursors of ABA lead to pre-harvest sprouting and photo-oxidation in rice. *Plant J.* **2008**, *54*, 177–189. [CrossRef]
18. Beale, S.I. Green genes gleaned. *Trends Plant Sci.* **2005**, *10*, 309–312. [CrossRef]
19. Bollivar, D.W. Recent advances in chlorophyll biosynthesis. *Photosynth. Res.* **2006**, *90*, 173–194. [CrossRef]
20. Qiu, N.; Jiang, D.; Wang, X.; Wang, B.; Zhou, F. Advances in the members and biosynthesis of chlorophyll family. *Photosynthetica* **2019**, *57*, 974–984. [CrossRef]
21. Zhang, H.; Liu, L.; Cai, M.; Zhu, S.; Zhao, J.; Zheng, T.; Xu, X.; Zeng, Z.; Niu, J.; Jiang, L.; et al. A point mutation of magnesium chelatase *OsCHLI* gene dampens the interaction between CHLI and CHLD subunits in rice. *Plant Mol. Biol. Rep.* **2015**, *33*, 1975–1987. [CrossRef]
22. Willows, R.D.; Gibson, L.C.; Kanangara, C.G.; Hunter, C.N.; von Wettstein, D. Three separate proteins constitute the magnesium chelatase of *Rhodobacter* sphaeroides. *Eur. J. Biochem.* **1996**, *15*, 438–443. [CrossRef]
23. Kannangara, C.G.; Vothknecht, U.C.; Hansson, M.; von Wettstein, D. Magnesium chelatase: Association with ribosomes and mutant complementation studies identify barley subunit Xantha-G as a functional counterpart of *Rhodobacter* subunit BchD. *Mol. Gen. Genet. MGG* **1997**, *18*, 85–92. [CrossRef]
24. Davison, P.A.; Schubert, H.L.; Reid, J.D.; Iorg, C.D.; Heroux, A.; Hill, C.P.; Hunter, C.N. Structural and biochemical characterization of Gun4 suggests a mechanism for its role in chlorophyll biosynthesis. *Biochemistry* **2005**, *31*, 7603–7612. [CrossRef]
25. Gibson, L.C.; Willows, R.D.; Kannangara, C.G.; von Wettstein, D.; Hunter, C.N. Magnesium-protoporphyrin chelatase of *Rhodobacter* sphaeroides: Reconstitution of activity by combining the products of the bchH, -I, and -D genes expressed in Escherichia coli. *Proc. Natl. Acad. Sci. USA* **1995**, *92*, 1941–1944. [CrossRef]
26. Sirijovski, N.; Olsson, U.; Lundqvist, J.; Al-Karadaghi, S.; Willows, R.D.; Hansson, M. ATPase activity associated with the magnesium chelatase H-subunit of the chlorophyll biosynthetic pathway is an artefact. *Biochem. J.* **2006**, *400*, 477–484. [CrossRef]

27. Gräfe, S.; Saluz, H.-P.; Grimm, B.; Hänel, F. Mg-chelatase of tobacco: The role of the subunit CHL D in the chelation step of protoporphyrin IX. *Proc. Natl. Acad. Sci. USA* **1999**, *96*, 1941–1946. [CrossRef]
28. Teng, Y.; Ye, L.; He, F.; Chu, W.; Wang, C.; Tan, Y. Expression Analysis of Key Genes of Chlorophyll Synthesis in A Yellow Leaves Mutant of Oryza sativa. *Chin. Agric. Sci. Bull.* **2017**, *33*, 30–35.
29. Lichtenthaler, H.K. [34] Chlorophylls and carotenoids: Pigments of photosynthetic biomembranes. *Methods Enzymol.* **1987**, *148*, 350–382.
30. He, F.; Shao, M.; Chen, T.; Wang, C.; Xu, X.; Tan, Y. Knockout of a magnesium chelating enzyme gene in rice using CRISPR-Cas9 system. *Fenzi Zhiwu Yuzhong (Mol. Plant Breed.)* **2019**, *17*, 5674–5680.
31. Pipitone, R.; Eicke, S.; Pfister, B.; Glauser, G.; Falconet, D.; Uwizeye, C.; Pralon, T.; Zeeman, S.C.; Kessler, F.; Demarsy, E. A multifaceted analysis reveals two distinct phases of chloroplast biogenesis during de-etiolation in *Arabidopsis*. *Elife* **2021**, *10*, e62709. [CrossRef]
32. Sandoval-Ibáñez, O.; Sharma, A.; Bykowski, M.; Borràs-Gas, G.; Behrendorff, J.B.Y.H.; Mellor, S.; Qvortrup, K.; Verdonk, J.C.; Bock, R.; Kowalewska, Ł.; et al. Curvature thylakoid 1 proteins modulate prolamellar body morphology and promote organized thylakoid biogenesis in *Arabidopsis thaliana*. *Proc. Natl. Acad. Sci. USA* **2021**, *118*, e2113934118. [CrossRef]
33. Allen, J.F.; de Paula, W.B.; Puthiyaveetil, S.; Nield, J. A structural phylogenetic map for chloroplast photosynthesis. *Trends Plant Sci.* **2011**, *16*, 645–655. [CrossRef] [PubMed]
34. Wang, Z.; Hong, X.; Hu, K.; Wang, Y.; Wang, X.; Du, S.; Li, Y.; Hu, D.; Cheng, K.; An, B.; et al. Impaired Magnesium Protoporphyrin IX Methyltransferase (ChlM) Impedes Chlorophyll Synthesis and Plant Growth in Rice. *Front Plant Sci.* **2017**, *8*, 1694. [CrossRef]
35. Sawicki, A.; Willows, R.D. S-adenosyl-L-methionine:magnesium-protoporphyrin IX O-methyltransferase from *Rhodobacter capsulatus*: Mechanistic insights and stimulation with phospholipids. *Biochem. J.* **2007**, *406*, 469–478. [CrossRef]
36. Adhikari, N.D.; Froehlich, J.E.; Strand, D.D.; Buck, S.M.; Kramer, D.M.; Larkin, R.M. GUN4-porphyrin complexes bind the ChlH/GUN5 subunit of Mg-Chelatase and promote chlorophyll biosynthesis in *Arabidopsis*. *Plant Cell* **2011**, *23*, 1449–1467. [CrossRef] [PubMed]
37. Farmer, D.A.; Brindley, A.A.; Hitchcock, A.; Jackson, P.J.; Johnson, B.; Dickman, M.J.; Hunter, C.N.; Reid, J.D.; Adams, N.B.P. The ChlD subunit links the motor and porphyrin binding subunits of magnesium chelatase. *Biochem. J.* **2019**, *476*, 1875–1887. [CrossRef] [PubMed]
38. Lundqvist, J.; Elmlund, H.; Wulff, R.P.; Berglund, L.; Elmlund, D.; Emanuelsson, C.; Hebert, H.; Willows, R.D.; Hansson, M.; Lindahl, M.; et al. ATP-induced conformational dynamics in the AAA+ motor unit of magnesium chelatase. *Structure* **2010**, *18*, 354–365. [CrossRef] [PubMed]
39. Jessop, M.; Felix, J.; Gutsche, I. AAA+ ATPases: Structural insertions under the magnifying glass. *Curr. Opin. Struct. Biol.* **2021**, *66*, 119–128. [CrossRef]
40. Miller, J.M.; Enemark, E.J. Fundamental characteristics of AAA+ protein family structure and function. *Archaea* **2016**, *2016*, 9294307. [CrossRef] [PubMed]
41. Koncz, C.; Mayerhofer, R.; Koncz-Kalman, Z.; Nawrath, C.; Reiss, B.; Redei, G.P.; Schell, J. Isolation of a gene encoding a novel chloroplast protein by T-DNA tagging in *Arabidopsis thaliana*. *EMBO J.* **1990**, *9*, 1337–1346. [CrossRef] [PubMed]
42. Zhang, H.; Li, J.; Yoo, J.H.; Yoo, S.C.; Cho, S.H.; Koh, H.J.; Seo, H.S.; Paek, N.C. Rice *Chlorina-1* and *Chlorina-9* encode ChlD and ChlI subunits of Mg-chelatase, a key enzyme for chlorophyll synthesis and chloroplast development. *Plant Mol. Biol.* **2006**, *62*, 325–337. [CrossRef] [PubMed]

MDPI

St. Alban-Anlage 66

4052 Basel

Switzerland

www.mdpi.com

Agronomy Editorial Office

E-mail: agronomy@mdpi.com

www.mdpi.com/journal/agronomy